Digital Television Standardization and Strategies

For a listing of recent titles in the *Artech House Digital Audio and Video Library,* turn to the back of this book.

Digital Television Standardization and Strategies

Katharina Grimme

Artech House
Boston • London
www.artechhouse.com

Library of Congress Cataloging-in-Publication Data
Grimme, Katharina
    Digital television standardization and strategies/Katharina Grimme.
    p. cm. — (Artech House digital audio and video library)
    Includes bibliographical references and index.
    ISBN 1-58053-297-7 (alk. paper)
    1. Digital television.   2. Broadcasting.   3. Business planning.
    I. Title.   II. Series.
    TK6678 .G75 2001
    384.55'4—dc21   2001026674

British Library Cataloguing in Publication Data
Grimme, Katharina
    Digital television standardization and strategies.
    (Artech House digital audio and video technology library)
    1. Digital television   2. Digital television—Standards
    I. Title
    621.3'88

    ISBN   1-58053-297-7

Cover design by Igor Valdman

International Standard Book Number: 1-58053-297-7
Library of Congress Catalog Card Number: 2001026674

10 9 8 7 6 5 4 3 2 1

To my mother and Hans

Contents

Acknowledgments

The final content of this book reflects the contributions of a range of individuals to whom I am very grateful. The work was originally submitted as a Ph.D. thesis at the University of Sussex, United Kingdom, and I especially want to thank Professor Alan Cawson for his guidance throughout the time I was working on my thesis.

I am indebted to many persons—too numerous to mention—who took the time to share their valuable knowledge with me. In particular, I would like to thank Jean-Pierre Coustel, Jörn Kruse, William Schreiber, and David Levy for helpful feedback. My gratitude also goes to Peter Kürble and colleagues at the Wissenschaftliches Institut für Kommunikationsdienste and to the team at the Paris-based think tank Prométhée, all of whom kindly supported my research in Germany and France.

Last but not least, I extend a special thank you to Ian for his help and patience and to my mother and Hans for their continuous encouragement and moral support.

1

Introduction

Developments in information and communications technologies (ICTs) in recent decades have been characterized by a process of accelerating convergence between formerly separate sectors such as media, telecommunications, and microelectronics. This change has come about through rapid advancements in the deployment of digital technologies for information transmission. Integrated digital systems (networks) are now an established means for the transmission and processing of information. Such systems provide numerous information services and form the basis for the emerging information society.

Integrated ICT networks are characterized by an interdependence of different system components that require some degree of compatibility. The services provided also introduce an interdependence among consumers, since the value of a network increases as more consumers join it. This effect is known as network externality (or network effect). In markets with network effects, compatibility is essential to establish a network as large as possible. Moreover, it has to be ensured that the components are interoperable. The way in which compatibility can be achieved, and the standardization process in particular, has received much attention from economists, business players, and policy makers. Part I of this book analyzes the characteristics of network markets and assesses different standardization approaches.

The speed at which new, integrated systems are developed and diffused is considered crucial for the competitiveness of companies, industrial sectors, nation states, and regions. The success of new networks depends essentially on three factors. First, network success is determined through the procedure through which an adequate level of compatibility is achieved. This is a result of the standardization approach. Second, the market entry strategies of systems suppliers have to be devised so as to provide value and adoption incentives to users. Finally, public authorities influence both the standardization process and technology adoption through their policy approach and regulatory framework.

This book analyzes the standardization and technology adoption of digital broadcasting. It looks at the most recent developments in the European, U.S., and Japanese audiovisual sectors.

The broadcasting industry is highly relevant for several reasons. Most importantly, the audiovisual sector is characterized by network externalities and compatibility requirements: Standardization issues are crucial. In addition, this industry sector has some distinct characteristics that make the development of advanced television (ATV) technologies a particularly important issue to industry players and society at large.

First, the emergence of ATV creates major strategic opportunities for several industrial sectors. These include consumer electronics and equipment manufacturers, network operators, broadcasters, and content producers—as well as, perhaps, formerly distinct sectors such as the computer and telecommunications industries and Internet service providers. A wide spectrum of business interests is represented, implying new challenges in both competition and cooperation. This is set against a background of high development costs [research and development (R&D) investments and initial set-up costs] and uncertain returns to investors.

Second, the domain of television has always been highly regulated on a national level because of its universal reach and strong influence on public opinion. A central objective of broadcast policy is to secure variety, which means not only maximizing consumer choice but also ensuring a plurality of opinion and values. Since the 1980s, however, the process of deregulation has gathered speed. This has resulted in a proliferation of service offers, an increase in cross-border operations, and a consolidation of commercial broadcasting markets centered around a few dominant companies. ATV technologies create new challenges for policy makers and regulators, but also provide an opportunity to reshape those market structures. This already turbulent environment is further challenged by the fact that television impinges on other sectors, often resulting in conflicting policy decisions. Forecasting future developments, therefore, has become increasingly difficult.

Third, the evolution of new television systems is causing yet another race in technology and performance between the Triad economies (Europe, United States, Japan). ATV technology is expected to revive the saturated television market that still remains the most lucrative for consumer electronics industries in Europe, Japan, and the United States.

Finally, the standardization of analog TV technology [for color TV and high-definition television (HDTV)] has been widely criticized as having led to an inefficient outcome. This has had a major influence on the approach adopted for digital television technology in the early 1990s. Part II analyzes the standardization process for digital television in Europe in the early 1990s. In comparing this process with the earlier attempts for HDTV, it is revealed that the open, committee-led Digital Video Broadcasting (DVB) Group, which bases decisions on consensus among its members, has enjoyed much more success than earlier standardization approaches.

The relative success of a standard-setting process is also reflected in subsequent technology adoption. Part III looks at the diffusion of digital TV technology in three European countries: France, Germany, and the United Kingdom. These are the largest broadcasting markets within the European Union (EU), and their respective market players have been the main drivers behind the development and standardization of digital-TV technology. This book carefully compares market entry strategies and efforts at standardization. To assess European developments in the aggregate, Part IV analyzes the U.S. and Japanese markets.

The analysis reveals that the European standardization effort has been an immense success and that the basic digital transmission standards developed by the DVB group are now being employed not only in Europe, but in dozens of countries worldwide. This, however, does not guarantee profitability for the involved companies. Business strategies to diffuse the new technology are equally important. Thus, we see widely differing outcomes depending on existing market structures and strategies adopted by national broadcasters. This book gives important insights into the issues that industry players have to consider when devising business strategies for the provision of digital-TV services.

PART

I

Network markets and standardization

2

Contents

Characteristics of network markets and their impact on corporate strategies

2.1 Definition of network markets

Today's ICTs offer a wide variety of products and services. Many of them consist of complementary components. The components have little or no value in isolation but generate value when combined with others. For the audiovisual sector, examples are home video components: The "hardware" such as the TV set, stereo, and video recorder and the "software" (TV programs, CDs, or videotapes) together provide entertainment services. Markets that are characterized by the interdependence of different components are referred to as systems markets. The economics of such systems markets with complementary products is distinct from that of individual products and merits special consideration.

These markets are characterized by network effects, meaning that the value of a

7

product or service to one user depends on how many other users there are [1]. Since this book discusses markets with both characteristics, the terms *systems market* and *network market* are used synonymously.

Based on economics literature, this chapter discusses the peculiarities of network markets and their impact on technology adoption and corporate strategies. Of particular importance is the need for coordination among the component suppliers, the expectations of consumers as to the development of the network, and supplier decisions with regard to compatibility.

2.2 Network externalities and positive feedback

Network markets are characterized by network externalities (or network effects). Externalities arise when one market *participant* affects others without compensating them. Externalities can be positive or negative. Examples for negative externalities are found in the ecological environment: If one country significantly pollutes the air, the neighboring country is affected as well. In information technologies, network externalities are usually positive. Communications networks, such as telephone, fax, and e-mail, are prime examples. The more users are connected to the network, the higher the value to one user because he or she can communicate with more people. This phenomenon constitutes a *direct* network effect.

Network effects also appear in *indirect* form, in what can be called *virtual networks*, such as a network comprising users of the same technology (e.g., IBM personal computers) [2]. Here, each user benefits from a larger network because the larger network facilitates the exchange of electronic files and makes a greater variety—or a higher quality—of software programs available. These indirect network effects occur as a result of demand-side economies of scale, where users value a network because it is widely used. The more users a network has, the higher is the demand to join the network—and the more willing are consumers to pay for joining. This type of network effect is applicable to television markets. Viewers do not communicate directly with each other, but, as members of a network, they can potentially benefit from a larger variety of content (programs).

In addition to demand-side economies of scale, there are also supply-side economies of scale (i.e., lower unit costs in the production of the network components). Supply-side economies of scale are, however, not specific to network markets but appear in almost every industry to a certain extent. These traditional economies of scale, based on manufacturing, are limited (i.e., they are generally exhausted at scales well below total market dominance). Demand-side economies of scale, in contrast, do not dissipate when the market grows larger. Hence, they tend to lead to a market-dominant position or, in the extreme case, to a monopoly situation.

The value of a network with demand-side economies of scale follows what is called *Metcalfe's law*, named after Bob Metcalfe, the inventor of the Ethernet system. According to this law, if there are n people in a network, and the value of the network to each of them is proportional to the number of other users, then the total value of the network (to all the users) is proportional to $n(n - 1) = n^2 - n$. This means that the value of a network increases exponentially with the number of users [3]. The larger a network is, the more value is gained by each additional user. Expressed differently, the larger a network is, the more attractive it becomes for new users to join this network.

This phenomenon is referred to as *positive feedback*: As the installed base of users grows, more and more users find adoption worthwhile [4]. Positive feedback means that the success of a particular system will reinforce itself. The competition between IBM and Apple computers illustrates this: As the IBM share of the PC market grew, users found this system more and more attractive. For Apple, the "positive feedback" was actually negative, the inevitable flipside of the coin being that the weaker got even weaker.

This forms an example of a market that "tips": In a market with strong positive feedback where two or more incompatible systems compete, only one may emerge as the winner (i.e., the market tips in favor of one player). It is unlikely that all can survive, although it is possible that others "coexist," often serving niche markets. Apple computers, for example, still exist but with a marginal market share.

Another example of a company benefiting from positive feedback is Nintendo. In 1985, it entered the U.S. market for home video games, which at the time was saturated and dominated by Atari. The new Nintendo system became highly popular, inducing a massive expansion of software and increasing the value of the network to consumers. Nintendo managed to make use of the positive feedback while at the same time retaining control over its technology. Every independent games developer paid royalties to Nintendo and had to promise not to provide its games for rival systems for two years following their release [5].

It has to be noted that positive feedback effects may not immediately generate winners. Incompatible systems can go head-to-head for a long period of time, or the losing network may simply stabilize with a much smaller market share. Consumer heterogeneity and product differentiation tend to limit tipping and sustain multiple networks. If the rival systems have distinct features sought by certain consumers, two or more systems may be able to survive by catering to consumers who care more about product attributes than network size.

A market with multiple, incompatible products over a long period of time reflects the value consumers place on variety. Enforced standardization could result in value losses. In other cases, any loss of variety seems a minor price to pay to achieve compatibility [6]. Questions that need to be answered in each case include the following:

 › Is heterogeneity large enough to enable more than one system to survive in the long run?

 › How do the benefits of network size compare to the social value of variety?

The uncertainties of technological progress also suggest that a certain value should be placed on diversity. In the presence of network effects, it can be difficult and costly to switch horses in midstream. A topical example is the Japanese HDTV system that is now widely regarded as inferior to the all-digital systems being introduced in the United States and Europe (see Chapter 11). Nippon Hoso Kyokai (NHK) and other Japanese suppliers did not expect a workable all-digital system to be feasible before the turn of the century, so they focused their efforts on a hybrid system. By promoting a single standardized system they are now poorly placed in the digital environment.

2.3   Lock-in and switching costs

The concept of "lock-in" arises whenever users invest in multiple, complementary, and durable assets that are specific to a particular information technology. Two such systems are incompatible if it is difficult or impossible to interchange components. Consumers who join a network invest in durable hardware, complementary software, and learning time. All these can be regarded as sunk costs and constitute a lock-in to a particular system. Consumers would need to write off such investments and undertake further expenditure if they switched to another system. These are known as switching costs [7].

An example would be a consumer who is contemplating a switch between vinyl and compact disc (CD) formats for music reproduction. Sunk costs include the record player and record collection. Switching would involve both investments in new hardware (CD player) and the complementary software (CDs). Learning costs in this example are minimal, but in situations such as alternative PC operating systems they could be foreseen as quite substantial. To convince consumers to switch, the new system has to be perceived as still adding value after switching costs are deducted. The CD system has become successful only because it offers significant improvements in convenience, durability, and sound quality over vinyl records.

New network systems that introduce a new, incompatible technology face difficulties due to network externalities and switching costs. Small-scale networks may necessitate a long payback period. In the introductory phase, while the network is small, its value to users and its attractiveness to new network members are low. The challenge for companies seeking to introduce such a network is to quickly win a certain number of users so as to create an installed base that will trigger positive feedback. Subsequently, due to both demand- and supply-side economies of scale, the value of the network will be higher and increasing rapidly.

The situation becomes even more difficult if more than one incompatible network is introduced. The new networks have to overcome both switching costs and competitive pressures with regard to market tipping. Establishing an installed base quickly will prove critical to survival.

2.4 First and second movers

The particular characteristics of network markets have implications for business strategies. In particular, the time of market entry relative to competitors can have a significant impact. In such markets, it is highly probable that first movers will be able to secure long-term competitive advantage, even a monopoly position [8]. The initial advantages for a first mover are as follows. Assuming that consumers differentiate the product from a general class of near-substitutes (e.g., digital TV versus other home entertainment activities) the company that first enters will exert market dominance. Once a network of consumers has been developed, economies of scale in all aspects of the business can be utilized. This should enable a firm to transform an initial advantage into a long-term position of power with the possibility of establishing barriers to entry against potential competitors.

In the event that first and second movers offer incompatible technologies, consumers will face switching costs as discussed in Section 2.3. Such a situation is likely to consolidate the position of the first mover since a new entrant will need to offer a very competitive product to induce consumers to make the switch. Entrants could engage in price/quality competition, loyalty schemes, or strong after-sales management to attract customers. Nevertheless, the fundamental criterion must still hold: The added net value of switching costs must be positive. This can more easily be achieved in markets with rapid technological progress since later movers are likely to be able to offer state-of-the-art products or services.

A skillful second mover may be able to take advantage of an infrastructure and technology reputation that the first mover has created. Moreover, a second mover can learn from the first mover's mistakes made with regard to market penetration, particularly those concerning issues such as consumer perceptions, distribution, and, if applicable, the regulatory environment. Firms with a tendency toward risk taking and with pioneering visions will normally be first movers, whereas those with strengths in production and marketing will tend to wait until the uncertainty over market development is reduced. This section's discussion illustrates that first and second mover advantages must be supplemented with appropriate business strategies. Section 2.5 expands on a number of the key issues.

2.5 Corporate strategies in network markets

This section considers the subsequent stages of market development where two or more incompatible systems are competing for revenues. In order to succeed, companies have to convince consumers to adopt their system as early as possible. Strategies to achieve this revolve around promoting a company's commitment to the future availability of the system's components. This section outlines key strategic approaches.

2.5.1 Consumer expectations

The question of which system evolves as the market leader will be strongly determined by consumer expectations. If consumers knowingly purchase "lock-in" hardware, it is reasonable to assume that they have given some thought to the complementary software, particularly its future price and availability. For example, it is reasonable to assert that sales of wide-screen TV sets will be strongly determined by current and future availability (transmission) of wide-screen programs.

Firms can give assurances to consumers on the future availability and price of system components or software, but these need to be seen as credible. Inevitably, after consumers are locked in, dominant firms may be tempted to renege on commitments. In the case of two competing systems it is consumer expectations at the turning point (i.e., the point where the market "tips") that are crucial. This is also known as achieving a critical mass and implies that each network has the same chance that the market will tip in its favor. Marketing strategies will prove decisive in network markets that are prone to tipping. Consumers need to be convinced that the network will survive and expand, thus increasing the flow of benefits over time.

Incumbent firms must also think carefully about the impact of future technological developments on their products, particularly as they concern their products' market price and attributes. Inducements may be needed to encourage consumers to invest today as opposed to postponing their decision in the hope that technological progress lowers prices [9].

2.5.2 Vertical integration

Consumers may perceive the vertical integration of hardware and software production as a commitment to the network. Alliances and joint ventures such as those currently observed in multimedia systems markets, linking cable TV and telephone companies together with consumer electronics firms, may go some way in reassuring consumers that they will not end up being stranded with useless systems. Full ownership and integration are also likely to lead to a greater willingness to invest in network growth since future rents are more easily captured. Identifiable and enforceable property rights are crucial in this regard.

Rewards to innovation such as patents or exclusive licensing arrangements will encourage the development of network markets. The difficulty, however, lies in balancing the benefits to society from innovation with the social losses induced during the period of monopoly rent maximization. It is here that we enter the regulatory domain. Regulators need to ensure that a vertically integrated firm, particularly a first mover, does not erect anticompetitive barriers. Control over hardware, for example, could be used to raise a rival's cost of software. In this way, a company could foreclose the market for content through a dominant position in hardware. This phenomenon is called vertical foreclosure. As ever, the objective of regulators is to find the right balance between technological progress and adverse use of market power.

This book argues that, in the case of digital television, network externalities are low; hence market foreclosure is unlikely to be relevant. For this reason, the book does not detail problems stemming from vertical foreclosure.

2.5.3 Pricing and equipment commitments

If consumers perceive that a firm may become dominant in the market, they may be reluctant to join a network for fear of future price rises. First movers will need to provide a credible pricing commitment on features of the system such as software. Firms can indirectly commit themselves to competitive software pricing by opening the market to independent suppliers. IBM, for example, encouraged independent software developers to write compatible programs when it was first introducing PCs.

Large initial investment in hardware and software is another way companies can show their commitment to the network. In this strategy, a company voluntarily reduces its freedom of action in order to illustrate its determination to succeed.

2.5.4 Reputation and advertising

An important factor in assuring consumers of the future availability and quality of software is the supplier's reputation. If consumers perceive that a firm will act to preserve its reputation, they will raise their expectations about the future network size and the availability of software [10]. Companies that command reasonable respect in other product or service markets, be they interrelated or not, will find it easier to attract consumers to a network. Unknown firms will require distinct and credible marketing.

Network markets require a high investment in consumer awareness. Ideally, firms must be able to convey the value their network provides through a simple message that appeals to a wide range of consumers. Advertising campaigns stressing the technological merits of hardware, therefore, are unlikely to be successful. It is ultimately the software that generates value, and this must be the initial focus. If the perceived utility of the software is high, the purchasing decision for the hardware is likely to be less elongated. Advanced publicity of service offerings enhances consumer confidence in the future development of the network and may discourage potential competitors.

2.5.5 Penetration pricing

Low prices in the introductory phase enable a firm to gain network users more quickly as the initial investment for the consumer is reduced. Subsidizing hardware should facilitate market penetration of a network system; hence, value, in the form of a diverse range of software, can be delivered more rapidly. Software vendors will profit in the long run from the low initial cost of hardware: If many users are acquired rapidly, software firms will benefit from high software sales. Hence, they may be expected to finance part of the initial hardware subsidy. Hardware subsidies do not require vertical integration in the industry. In the cellular telephone sector, for example, handsets are heavily subsidized by network operators to attract subscribers.

Hardware rental constitutes another form of subsidy. By renting hardware, consumers can be more easily persuaded to adopt a new system because initial expenditures are lower. The risk of being "stranded," in the event that the chosen system loses in the race to become the definitive technology, is also reduced.

Allowing the rental of hardware involves a trade-off for network sponsors. On the one hand, an installed base should be achieved more rapidly since consumers will no longer fear the possibility of lock-in. An installed base contributes to the determination of the dominant standard in the market. Conversely, since no lock-in effect is created future revenues may be uncertain. Dissatisfied consumers may transfer to another network since switching costs are much lower. Network sponsors are more likely to consider using the hardware rental strategy if they can identify a definitive competitive advantage within their software offerings. In general, however, this requires invoking assumptions about consumer preferences that may prove unrepresentative.

2.6 The impact of compatibility on competition

In view of the high stakes available, competition in network markets is likely to be aggressive. If the market can be expected to tip in favor of a single system, firms are effectively bidding for a long-term stream of monopoly profits. This is the prize after the initial period of losses resulting from the large initial investments into the market. Uncertainty over financial viability is heightened if a firm decides to introduce an incompatible network system. A firm will need to make a careful cost-benefit analysis when making a decision on compatibility. A company unable to finance the investments required in the initial period is not likely to opt for an incompatible system.

Furthermore, a compatibility choice will depend on the perceived competitive advantage of a firm. In the case of compatible systems, competition is not between the overall package (including the network size), but between the specific cost and performance characteristics of each component individually. This means that if one firm has a distinctly superior overall package, including its product and service offerings, its installed base, and its reputation, that firm is likely to prefer incompatibility and may spend resources to block compatibility. In contrast, if one firm has a distinctly superior component, which can be complemented by another, again distinctly superior, component of a second firm, both firms are likely to prefer compatibility and may spend resources to achieve it.

Compatibility effects on pricing competition over time have been analyzed by Katz and Shapiro who found that compatibility relaxes competition early in the product life cycle because the threat of tipping is reduced or nonexistent [11]. However, because compatibility prevents one firm from gaining control of the market, competition tends to intensify later in the product life cycle. Firms that are confident that the market will tip in their favor are likely to oppose compatibility and opt for proprietary technologies.

Asymmetries in the compatibility choices of firms are especially likely when one of the firms is an entrant and the other an incumbent. A new entrant with an incompatible technology may find it particularly difficult to establish an installed base and build reputation; hence it may opt for compatibility. Only when the technology is distinctly superior and the benefits easily conveyed can a new entrant be confident that it can succeed independently. An incumbent is likely to perceive its expertise and existing structures as a competitive advantage and may therefore choose to introduce an incompatible system.

Compatibility may be introduced exogenously in the form of standardization and regulation. It should be noted that the literature on network markets tends to use the terms *compatibility* and *standardization* synonymously, but a distinction is applicable. Standardization describes *durable* compatibility (i.e., compatibility fixed for the longer term as a result of a standardization process). Chapter 3 examines the different ways in which such a standardization process may take place.

2.7 Regulation in network markets

Positive feedback effects based on network effects pose challenges to regulators in two ways. As noted previously, the emergence of a dominant firm may give cause for concern with respect to consumer welfare. Beggs and Klemperer, for example, provide an analysis showing how a firm with a large installed base will tend to set higher prices [12]. This forms an indication that competition is distorted. Regulation should aim to ensure that the market for provision of network services is not foreclosed. It might be wrong, however, to ban the introduction of a new system on the grounds that it is likely to evolve into a monopoly. Because of significant economies of scale in network markets, it may be most efficient if one or only a few companies supply the entire market. Hence, a monopolistic or oligopolistic market structure may be in the best interest of consumers.

Regulatory intervention may also be required if rival companies agree to cooperate to establish compatibility, and this then evolves into a collusive arrangement with anticompetitive interests. In high-technology markets where sharing and managing risk are paramount, regulators will again need to be sensitive and balance the social gains from R&D against the losses associated with market foreclosure.

The issue of lock-in creates further problems due to the complementary nature of ICT products within networked systems. Changing any single component can be very costly for users. The lock-in that results will increase the competitive advantage for any firm in the market. In formulating an adequate competition policy, the first task will be to decide on what constitutes a reasonable representation of a competitive network market. Unless some basic indicators are agreed upon, any assessment of the desirability of a dominant market position is likely to be quite arbitrary. In rapid-growth ICT markets, periods of market advantage can be quickly eroded by new products and services; hence intervention may not be necessary.

Proactive regulation is more justified on matters such as mergers and acquisitions, predatory pricing, retailing of products through exclusive dealerships, or "bundling" of products and services. A current example of the latter is the landmark trial against Microsoft, which has been accused of abusing market power over the Windows operating system to suppress competition and innovation across the industry. In particular, the case concerns the tying of Internet software (i.e., the Explorer browser) to the

Windows system. Interestingly, the Microsoft defense argues that its high market share (around 90% of the global PC market) does not constitute monopoly power and that the market remains open to the development of alternative system software. The task for the authorities is twofold. First, it is necessary to decide if these components are separate products or if they form an integrated operating system. Second, it is necessary to analyze whether consumers would indeed benefit from a greater integration of the browser and operating system.

Industry-specific regulation is often used in oligopolistic markets such as telecommunications and utilities. Highly concentrated markets such as these are subject to, among other things, pricing, quality, interconnection, and entry requirements since it is assumed that temporary advantages are less likely to be eroded by technological change. To deal with those issues, specific regulatory agencies are often established. The principal problems with these are the social losses induced by large bureaucracies and the fact that bureaucrats are unlikely to respond adequately to the dynamic pressures within network markets. This could lead to market distortions or even the stagnation of technological development.

In summary, it can be stated that the type of regulatory approach chosen has an impact on how market structures evolve in network markets. Therefore, companies in these markets have to take regulatory activities into consideration when devising their corporate strategies.

Endnotes

[1]   Network effects were first defined and examined by Rohlfs, J., "A Theory of Interdependent Demand for a Communications Service," *Bell Journal of Economics* 5, No. 1, 1974, pp. 16–37. However, their significance for the economy and for corporate strategies was only recognized from 1985. An overview of existing work is provided by Katz, M. L., and C. Shapiro, "Systems Competition and Network Effects," *Journal of Economic Perspectives*, Vol. 8, No. 2, 1994, pp. 93–115.

[2]   This distinction originates from Katz, M. L., and C. Shapiro, "Network Externalities, Competition and Compatibility," *American Economic Review*, Vol. 75, 1985, pp. 424–440.

[3]   Shapiro, C., and H. Varian, *Information Rules, a Strategic Guide to the Network Economy*, Boston, MA: Harvard Business School Press, 1998, p. 184.

[4] The significance of positive feedback was first emphasized by Arthur, B. W., "Competing Technologies, Increasing Returns, and Lock-in by Historical Events," *Economic Journal,* Vol. 99, 1989, pp. 116–131.

[5] Shapiro, C., and H. Varian, *Information Rules, a Strategic Guide to the Network Economy,* Boston, MA: Harvard Business School Press, 1998, p. 178.

[6] This trade-off between standardization and variety is discussed in Farrell, J., and G. Saloner, "Standardization and Variety," *Economic Letters,* Vol. 20, 1986, pp. 71–74.

[7] For an overview of literature on switching costs, see Klemperer, P., "Competition When Consumers Have Switching Costs: An Overview with Applications to Industrial Organization, Macroeconomics and International Trade," *Review of Economics Studies,* Vol. 62, No. 4, 1995, pp. 515–539.

[8] On first mover advantages, see Lieberman, M. B., and D. B. Montgomery, "First Mover Advantages," *Strategic Management Journal,* Vol. 9, 1988, pp. 41–58.

[9] Katz, M. L., and C. Shapiro, "Systems Competition and Network Effects," *Journal of Economic Perspectives,* Vol. 8, No. 2, 1994, pp. 93–115.

[10] See Katz, M. L., and C. Shapiro, "Systems Competition and Network Effects," *Journal of Economic Perspectives,* Vol. 8, No. 2, 1994, pp. 103–105.

[11] Katz, M. L., and C. Shapiro, "Product Compatibility Choice in a Market with Technological Progress," *Oxford Economic Papers,* "Special Issue on the New Industrial Economics," November 1986, pp. 146–165.

[12] Beggs, A., and P. Klemperer, "Multiperiod Competition with Switching Costs," *Econometrica,* Vol. 60, No. 3, 1992, pp. 651–666.

CHAPTER

3

Approaches to standardization

3.1 Definition of compatibility standards

A workable definition of a basic standard is offered by David and Greenstein: "a set of technical specifications adhered to by a producer, either tacitly or as a result of a formal agreement" [1]. They also distinguish between reference, minimum quality, and compatibility standards. The former two provide signals that a given product conforms to the content and level of certain defined characteristics. These standards serve to reduce the transaction costs of users who are in the process of evaluating a range of technological specifications. This chapter will focus on compatibility standards. Compatibility standards assure the user that an intermediate product or component can successfully be incorporated into a larger system that comprises closely specified inputs and outputs. A product that conforms to a compatibility standard can serve as a subsystem within a larger system provided by a range of suppliers. An example of such a standard is the compatibility of the different components of a hi-fi system where the amplifier, the music medium (e.g., CD player), and the speakers have to be designed in such a way that they can be connected and will work together.

3.2   Advantages and disadvantages of standardization

Classic examples of failures to attain compatibility include fire hoses that do not fit into fire hydrants, railroad cars that do not match railroad tracks, or simply people who speak different languages. It is misleading, however, to view incompatibility simply as a standardization failure. Although standardization has obvious benefits, obtaining and maintaining compatibility also involves costs. As discussed in Section 2.7, in its decision on compatibility, a firm has to weigh the costs and benefits of standardization.

3.2.1   Benefits of standardization

In network markets with complementary components, the benefits of standardization to component suppliers are ultimately due to economies of scale on both the demand and supply sides. When the components of different suppliers are interchangeable, there may be greater opportunities to exploit these together with learning effects and technological spillovers in the development and production of components.

Software products, such as TV programs, computer software, or any other information goods, are characterized by high development costs (fixed and sunk costs) and nearly constant marginal costs [2]. In broadcasting, for example, total cost for program production is independent of the audience size. Once the content is produced, the cost of multiplying and distributing it to one more viewer (i.e., the marginal cost) is near zero. This results in strongly declining average costs with increasing viewers. Hence, if the hardware (TV technology) is standardized, the economies of scale are substantial. Furthermore, standardization facilitates a reduction of transaction costs, especially those associated with information and learning. ICT networks often require a substantial initial learning effort. With a standardized system, users only have to learn to operate one system rather than a range of different systems. Standardization also enhances variety by allowing consumers to mix and match (differentiated) components from various suppliers. With an increasing number of network participants the number of differentiated system components (e.g., software and receiver equipment) increases and thus provides a larger choice and value to the user. Hence, reduced variety on the system level leads to more variety on the component level. Moreover, the earlier a standard is set, the faster any unnecessary research can be halted. This releases resources for the improvement and differentiation of components in the standardized system and should contribute to increased variety and consumer choice. Finally, in a standardized network, consumers need not fear that the technology they have chosen will end up being the loser, leaving them stranded. This certainty reduces consumer reluctance to joining a network and is thus likely to accelerate the adoption of a new technology.

3.2.2 Potential disadvantages of standardization

There are also a number of disadvantages of standardization. The principal problem is reduced product variety. Consumers have fewer differentiated products to pick from, especially if standardization prevents the development of promising but unique and incompatible new systems [3]. This is likely to result in welfare losses if product variety is regarded as valuable. Consumer preferences may vary according to user requirements (horizontal preference differentiation) or a differing preparedness to pay for a certain level of quality (vertical preference differentiation) [4].

A limiting effect on future innovation is often cited as another disadvantage of standardization. The argument is that standards can limit product development since incompatible products are excluded from the market. Furthermore, to maintain the benefits of a conventional product, it is often mandated that a new standard be compatible with the former one. This demand for compatibility over time is particularly relevant in markets for high-cost consumer durables, such as TV sets. Here again, incompatible innovations are often excluded, implying that new standards, which comply with existing ones, may be technologically inefficient. This could even occur over several product generations. In this way, the first standard influences the future development of the product even in the long run [5]. It is debatable, however, whether standardization inhibits innovation because certain research areas are abandoned, or whether it supports innovation because more R&D resources can be devoted to potentially successful product lines.

3.2.3 Levels of standardization

Also of crucial importance is the level of standardization. In most standardization processes it is not just a discrete "yes or no" decision, but rather a judgment as to the appropriate level of standardization. If, in addition to basic functions, further detailed features of a technology are to be taken into account in the standardization process, a trade-off occurs: A high level of standardization guarantees improved compatibility but may incur excessive costs. These costs include the standardization procedure itself, losses in product variety, and the possible impediment of technological change. The level of standardization also has an impact on product variety. A lower level of standardization maintains product variety.

Less than full standardization also holds the advantage that a certain degree of competition can be upheld, allowing different suppliers to offer differentiated products with proprietary elements. Despite the use of basic standards, detailed product features can vary, thus allowing firms to differentiate their products from those of competitors. Competition on the market can avoid a cartelization and generally implies lower costs for consumers.

In summary, standardization is favorable in markets characterized by strong network effects, economies of scale, and significant transaction costs. It is less desirable if consumers exhibit wide differences in preferences over the services the network provides. In general, the crucial challenges are to decide upon the technologically optimal standard and then proceed to introduce the right degree of standardization.

The next question to be analyzed is therefore which standardization approach is best suited to determine the technologically optimal standard and, at the same time, is most likely to determine the right degree of standardization. Section 3.3 discusses this issue.

3.3 Approaches to standardization

Standards always require a certain level of coordination between different suppliers. The efforts to achieve this coordination comprise a standardization process. Such processes can be distinguished in four basic types: sponsored and unsponsored standards, standards committees, and mandated standards [6]. "Sponsoring" is defined as the willingness to make investments in the form of penetration pricing to establish the technology, with the intention of gaining additional profits in a later phase. The sponsor of a network is the firm or group of firms that creates and operates the network. The first two types of standardization evolve from a market-driven approach and are referred to as de facto standards.

In the committee and the mandated approach, in contrast, explicit, organized coordination leads to the choice of a standard. These two latter forms are referred to as de jure standards although only in the last case are they formally enforced by law. Each type of standardization is considered in more detail in Sections 3.3.1 through 3.3.3.

3.3.1 The market-driven approach

Sponsored and unsponsored standardization approaches are market-driven. This section explains and assesses these concepts.

3.3.1.1 Unsponsored standards

Unsponsored standards have no identified originator holding a proprietary interest, nor any subsequent sponsoring agency, but they nevertheless exist in a documented form in the public domain. The initial conditions resemble a competitive market where no firm is able to affect market price with its output decisions. Unsponsored standardization can result, for example, when increasing returns to adoption exist or when users are "locked in" by historical events. A classical example of the latter is the adoption of the QWERTY keyboard layout [7]. The QWERTY arrangement of keys evolved due to technical constraints in the era of mechanical typewriters. It was necessary that those keys often used in immediate succession should be arranged far away from each other so as to avoid the jamming of the type bars. With more advanced technology (electronic typewriters and computers), this arrangement became unnecessary. Although it was proven that there were superior keyboard arrangements that would allow faster typing, namely the Dvorak simplified keyboard (DSK), users stayed locked in to the QWERTY layout. The switching costs (replacement of keyboards and retraining of users) were perceived to be higher than the potential benefits from faster typing. However, this argument has been questioned: Liebowitz and Margolis claim that the test by the U.S. Navy that allegedly proves the superiority of the DSK keyboard layout and forms the main argument for the inferiority of QWERTY was not objective as it was carried out by Dvorak himself—the inventor of the DSK keyboard [8].

In today's ICT markets unsponsored standards are seldom observed. Most newly emerging networks involve at least one sponsoring company. That is why literature on standardization, when referring to a market-driven, de facto approach to standardization, usually implies sponsored standards.

3.3.1.2 Sponsored standards

In the case of sponsored standards, one or more sponsoring entities hold a proprietary interest. These can be suppliers (e.g., the developer who holds a patent), users or private cooperative ventures comprising a number of firms. Depending on the market power of such firms, their vested interest can induce other firms to adopt a particular set of technological specifications. In sponsored standardization processes, proprietary control can create incentives for firms to manipulate technical standards so as to make their goods compatible with complementary components or substitute systems, as well as to engage in strategic price setting. Sponsoring agents need to anticipate the reaction of their rivals and plan accordingly. Well-targeted pricing strategies may facilitate lock-in to the sponsored standard. Competition between incompatible systems typically takes place in an oligopoly situation (or between very few technologies). The purest form of sponsored technologies is evident if the owners prevent other firms from using their technologies. An example of such a "battle of systems" was the rivalry between Sega and Nintendo in the video games market in the early 1990s.

Crucial to the outcome of sponsored standardization processes is the ownership of assets. This aspect is summarized well by David and Greenstein:

> Initial asset ownership conditions—often resulting from previous episodes of market competition—can influence the evolution of subsequent standards, because initial asset ownership gives an advantage in the design and production of related components of a developing system. Such advantages, in turn, determine the abilities of particular firms to take on leadership roles in the design of interrelated systems of components that become de facto standards [9].

This implies that incumbent firms, due to their experience and an existing reputation, may have an advantage in the standardization process. In particular, an existing dominant player can exert a decisive influence on the standardization process.

So far, it has implicitly been assumed that technical incompatibilities, once implemented, will persist, causing static inefficiencies. The analysis changes, however, with the possibility of "gateway technologies" (converters, translators, and emulators). These have the possibility to achieve compatibility at a later point in time. Gateway technologies can reduce the social cost of failures to ex ante standardization [10]. Furthermore, the anticipation of a later development of gateway technologies can have an influence on the initial standardization process. It can be envisaged that if gateway technologies are anticipated, initial incompatibility will be more acceptable to suppliers and consumers.

3.3.1.3 Critique of market-driven approaches

A fully market-driven approach to standardization might operate as follows. Considering a TV network, a number of sponsors would compete to develop TV systems and, at any given time, a system could be introduced. Once a leader has been established, sponsors of rival systems would introduce their own proprietary systems. Any sponsor commercializing its technology would choose the price for its receivers and transmission equipment, or put into place a licensing policy (e.g., at a specified royalty rate).

Leaving the choice of a technology to the market holds some advantages. In particular, the market can judge quality ex post and the market gets a firm indication of cost. Furthermore, bright ideas from unexpected sources may be able to thrive better in a market than under an administrative approach. Finally, competition is likely to keep prices low early on, which will help to establish a viable network [11].

The market mechanism for standardization has been shown to work efficiently if a new technology is introduced sequentially rather than simultaneously. If an important market player makes a unilateral public commitment to one standard, others will most likely follow, knowing that they will be compatible with the first mover, and probably also with later movers. This "bandwagon" mechanism can sometimes achieve rapid and effective coordination. One example is the adoption of a protocol for scrambling satellite signals to cable TV systems in the United States in the mid 1980s. Several alternatives were available but after the largest market player, Home Box Office, chose the VideoCipher standard, other users rapidly followed suit and coordination was quickly achieved [12]. Hence, under certain conditions a market-driven standardization approach can lead to an efficient outcome.

There are, however, arguments against the desirability of a market-driven approach. The above example of a de facto standard through the creation of a bandwagon momentum carries the risk that an inferior technology might be adopted. In addition, the natural monopoly aspect of technology standards may lead to reductions in consumer welfare.

On the other hand, if no dominant player sponsors a particular standard, adoption decisions by users may be postponed for fear of committing themselves to durable assets that may not gain wide acceptance or soon be rendered obsolete. Especially when switching costs are very high, this can delay the development of the market as a whole and leave it fragmented among technologically incompatible systems, thereby discouraging investment in complementary products and services [13]. Market competition may also leave some consumers with incompatible equipment once the market has tipped in favor of another system.

It is argued that many of the assumptions underlying the notion of efficient market equilibriums do not apply to technology choice problems, especially those with network externalities. Market forces cannot be expected to yield a socially desirable portfolio of development projects, since system sponsors decide whether to enter or exit from the development competition. It is also argued that there is no reason to expect that market forces will cause a new generation of products to be introduced at the optimal time. Furthermore, the market mechanism is not always successful in achieving standardization. Theoretical literature on standardization suggests that the market might not "tip" when there is no clear leader and when there are different preferences for standards [14].

Generally, it can be assumed that the involved companies will be able to decide if they want to engage in systems competition in which they may finally lose. If the risk appears too high and economies of scale are promising, they might initiate a cooperative approach to standardization. This type of approach is discussed in Section 3.3.3.

3.3.2 The direct regulatory (interventionist) approach

Governments can set standards via industry-specific regulatory bodies or official organizations. This would require producers to adopt certain technological specifications [15]. Such regulation has typically been applied in public service industries, such as telecommunications in the 1970s and 1980s, where governments determined and mandated national standards in great detail. The interest of government bodies to act as standards setters can result from a variety of factors [16]:

▸ The government perceives a service or market to exhibit "public" or "merit" good characteristics. In the absence of regulation, provision of services may not be socially optimal.

▸ Standardization may affect important national goals, such as the protection of domestic employment or competitiveness.

▸ The government concludes that market-led or committee standardization processes have undesirable outcomes (e.g., inhibit effective competition).

The main argument in favor of interventionist standardization is the certainty that a common standard is achieved. The principal counterargument is that regulators may not determine the most efficient standard. Indeed, in areas with rapid technological change regulatory authorities are likely to have a significant informational disadvantage. The case of HDTV standardization in Europe serves as a prime example for a public authority picking and promoting an unsuccessful technology (see Section 5.2).

A further argument against the direct regulatory approach is that other policy objectives influence the outcome. The most common example is the setting of standards so as to create strategic barriers to trade (nontariff barriers) for other countries. These can be permanent, to protect the national industry, or temporary, to provide a competitive advantage for a national innovator.

With an increasing focus on international competition, and having learned from past failures with respect to supporting "national champions," the standardization activities of government bodies have been progressively reduced during the past decade. It is often difficult for government bodies to assess whether a market failure is evident and what effects the mandating of standards would have on the industry. Ill-considered standardization efforts are likely to result in market distortions.

Government standardization bodies are working much more closely with industry and often work toward a consensus on standards through the formation of voluntary committees. Agreed standards are subsequently only "rubber-stamped"(i.e., formally mandated) by standardization authorities. An example of such a consensus-based approach is the DVB group, which will be discussed in detail in Section 5.3.

In the case of one dominant player that is able to trigger a bandwagon effect but that may not have chosen the most efficient technology, the situation may be different. As there will be no possibility for firms to initiate a cooperative approach, regulators may feel obliged to intervene. Again, it is questionable whether regulators would be best suited to mandate a standard. Instead of directly intervening, regulators could aim to initiate a committee approach. An alternative approach for regulators would be to make sure that the dominant player does not abuse its position by charging monopoly prices and creating entry barriers for potential competitors. Rather than an interventionist approach, this would represent effective competition policy.

3.3.3 The committee approach

Since the 1980s, official standard-setting bodies such as the European Telecommunications Standards Institute (ETSI), the Comitée Européen de Normalisation Électrotechnique (CENELEC), or the International Standards Organization (ISO), have been criticized for their slow and cumbersome response to calls for new technological standards [17]. Consequently, an increasing number of firms today are looking for alternative, private, industry-led modes of achieving rapid technological coordination. Their approach is analyzed in this section.

3.3.3.1 Characteristics of a committee

Private standards consortia aim to achieve consensus through a committee approach. These organizations range from user groups to trade organizations and vary considerably in terms of their structures, membership requirements, rules, and procedures. They can act as support and reinforcement to official standardization bodies, but more recently the trend has been toward the formation of committees seeking to bypass the formal process altogether.

The typical procedure of a committee begins by referencing an existing model that guides the technical context within which specific standards are developed. In the subsequent formal process, new components are developed. All feasible options are put down in a standard protocol. On the basis of this protocol, the options are discussed and tested until, usually in a consensual approach, a single common specification is agreed upon.

Committees can include representatives from industry and consumer groups. In the case of supranational committees, representatives of official national and international standardization bodies participate in the negotiations. Committees can either be established through influence of authorities or form as a result of voluntary cooperation between industry representatives.

In a committee approach, the involved parties negotiate, and agreements are reached through voluntary cooperation. The main idea behind a committee solution is that technological know-how can be collected and shared by the involved industry players. This facilitates a much better appreciation of technical issues and should make for more targeted standards.

3.3.3.2 Assessment of committee approaches

The committee form offers the advantages of flexibility and speed since its membership, internal organization, and procedures may all be tailored to the specific task at hand [18]. This is particularly important in information and communication technology where the pace of technical progress is swift and product cycles are short. Furthermore, standard-setting efforts in an industry committee are close in nature to collaborative R&D and may contribute to technological advancements.

Another advantage of private consortia is that rules and guidelines as well as political considerations do not distort the process. In addition, the resources that can be provided by a private committee will often far exceed those available to formal standardization bodies. A further main distinction of private consortia is the fact that strategic commercial considerations are already included in the development of new standards.

Private consortia also have disadvantages: Rival groups may duplicate one another's efforts; start-up costs may be high; and there may be a lack of specialized administrative experience. More importantly, social risk arises through the "privatization" of a sphere of national and international cooperation for the provision of public goods. These profit-driven ventures, in bypassing inefficient institutional procedures, also bypass important public interest safeguards built into the formal process. These include rules guaranteeing free and equal access to committee deliberations and public circulation of draft and requirements for nonexclusive licensing of proprietary technologies. Few industry committees are likely to meet the high standards of openness and due process that are maintained within formal bodies [19].

A further problem associated with private committees is the credibility of adherence to the standard adopted. Some consortia may be formed for strategic reasons that are transient in nature, or large consortia may not survive in the long term due to conflicting interests among members when further enhancements and applications are contemplated. It may be advantageous to combine a committee with an official regulatory approach in the sense that once standards are agreed on in the industry-led group they can then be formally mandated. In order to maximize the benefits of standardization, various arrangements combining formal bodies and private committees are possible. In this way, some of the public interest safeguards of the formal process can be maintained in conjunction with the benefits stemming from the higher efficiency of a private committee.

It has to be determined, however, whether the committee's standard offered for formal approval leaves sufficient discretion for the authorities. If the specification is offered as a fait accompli, perhaps even after the launch of products in which it is embodied, the role of the standard-setting body would be reduced to that of a "rubber-stamping" agency, and procedural safeguards are likely to be lost.

Further criticism of standardization committees is that decisions often become technically complex thereby rendering the decision-making process inaccessible to many others, in particular representatives of the user community [20]. Moreover, a committee process can be slow due to the procedures involved to enable a consensual approach. Finally, firms participating in a committee may "sit back" and not push innovation as they would if they had to compete under free market conditions.

In addition to these criticisms, the committee approach has other shortcomings. First of all, no agreement may be reached at all, although it is quite possible that social benefits could be maximized by nonstandardization. Second, a committee process involving a large number (or all) of the players in a given market may agree on a standard that establishes a cartelized market. This is particularly the case where a high level of standardization is achieved (i.e., standards are defined in great detail).

3.3.3.3   Determinants of the outcome of a committee process

With regard to the level of standardization, committee approaches are generally favorable. Blankart and Knieps suggest that committees are more likely to agree on lower levels of standards (i.e., basic specifications such as a physical connection within a network) [21]. With increasingly higher levels, firm-specific preferences are touched upon and agreement will be much harder to achieve. Firms will have a vested interest in keeping some proprietary elements to ensure future revenues. If these are not guaranteed, the committee process may fail to achieve standardization.

It is, therefore, likely that, in a committee process, firms will agree on a lower level of standardization. This has obvious benefits (see Section 3.2.3). In a bureaucratic process, in contrast, it can be expected that a high level of standardization will be adopted, resulting from regulators' objective to maximize their power. This argument derives from the insights of the public choice school of economics [22].

The outcome of a committee approach will depend on the organizational setup. This includes the composition of its members, the formal procedure by which a consensus is achieved (e.g., agreement by all members or sufficiency of a qualified majority), as well as the time frame set for the approval of a standard. The design, timing, and likelihood of the adoption of specifications for particular standards also depends largely on the following features:

▸ The degree to which the technology is already developed, its characteristics and cost structure, and whether these are known in the industry;

▸ The degree to which investments in particular standards have already occurred;

▸ The degree to which relevant decision makers influence the process, as well as the impact of international political issues on the group (e.g., industrial policy objectives of governments [23]);

▸ The degree to which powerful members (e.g., dominant incumbents) can influence the committee process through their strategic behavior.

The latter two points in particular illustrate that the regulatory framework, the impact of policy makers, and the existing market structures can affect the outcome of a committee process.

3.4 Concluding remarks

This chapter argues that in ICT markets that are characterized by network effects, compatibility arrived at through standardization can hold considerable advantages. The involved companies can best make the decision as to whether these advantages exceed the costs. Standardization can then proceed through a market-driven process or a committee approach. The direct regulatory approach is generally undesirable. If public authorities fear a market failure, they should aim to initiate a committee approach, rather than mandating a standard, while at the same time applying effective competition law.

Committee approaches require openness and an equality in the influence of all participants. Furthermore, the organizational setup has to safeguard public interests. At the same time, the procedures have to be designed in such a way that they do not burden product development.

Endnotes

[1] David, P., and S. Greenstein, "The Economics of Compatibility Standards: An Introduction to Recent Research," *Economics of Innovation and New Technology*, Vol. 1, 1990, p. 4.

[2] Information goods, as defined by Shapiro, C., and H. Varian, *Information Rules, a Strategic Guide to the Network Economy*, Boston, MA: Harvard Business School Press, 1998, p. 3, comprise all types of information, such as books, movies, stock quotes, and databases, that are of value to consumers.

[3] Katz, M. L., and C. Shapiro, "Systems Competition and Network Effects," *Journal of Economic Perspectives*, Vol. 8, No. 2, 1994, p. 110.

[4] See Woeckener, B., "Standardisierungspolitik für die Informationsgesellschaft," *Jahrbuch für Nationalökonomie und Statistik 215/3*, 1996, p. 260.

[5] For an overview of existing literature on this topic, see Thum, M., *Netzwerkeffekte, Standardisierung und Staatlicher Regulierungsbedarf*, Tübingen: J. C. B. Mohl (Paul Siebeck) Verlag, 1995, pp. 19–22. Heuser, U. J., *Hochauflösendes Fernsehen: Fallstudie und Analyse des Internationalen Standardisierungsprozesses*, Dissertation, Köln: Universität, 1992, pp. 9–10, discusses this in the context of the HDTV standardization process.

[6] These distinctions originate from Katz, M. L., and C. Shapiro, "Technology Adoption in the Presence of Network Externalities," *Journal of Political Economy*, No. 94, 1986, pp. 830ff., and, with slight variations, are now generally accepted in literature on standardization.

[7] For a detailed description see David, P., "Clio and the Economics of QWERTY," *American Economic Review* (Papers & Proceedings) 75, 1985, pp. 332–336; and David, P., "Understanding the Economics of QWERTY: The Necessity of History." In *Economic History and the Modern Economist*, pp. 30–49, Parker, W. N. (ed.), Oxford, England: Basil Blackwell, 1986.

[8] Liebowitz, S. J., and S. E. Margolis, "The Fable of the Keys," *Journal of Law and Economics*, No. 33, 1990, pp. 1–25. For a description of both sides, see Lucky, R., *Silicon Dreams: Information, Man, and Machine*, New York: St. Martin's Press, 1991.

[9] David, P., and S. Greenstein, "The Economics of Compatibility Standards: An Introduction to Recent Research," *Economics of Innovation and New Technology*, Vol. 1, 1990, p. 13.

[10] This is discussed, for example, by Braunstein, Y. M., and L. J. White, "Setting Technical Compatibility Standards: An Economic Analysis," *Antitrust Bulletin*, No. 30, 1985, pp. 337–356.

[11] See Farrell, J., and C. Shapiro, "Standard Setting in High-Definition Television," *Brookings Papers on Economic Activity: Microeconomics*, 1992, pp. 29–30.

[12] See Besen, S. M., and L. L. Johnson, *Compatibility Standards, Competition, and Innovation in the Broadcasting Industry*, California: The RAND Corporation, R-3453-NSF, 1986; and Farrell, J., and G. Saloner, "Coordination Through Committees and Markets," *RAND Journal of Economics*, Vol. 19, No. 2, Summer 1988, pp. 235–252.

[13] David, P. A., and M. Shurmer, "Formal Standard-Setting for Global Telecommunications and Information Services," *Telecommunications Policy* 20/10, special issue, 1996, pp. 801–802.

[14] For an overview of this literature see Besen, S. M., and G. Saloner, "Compatibility Standards and the Market for Telecommunications Services." In *Changing the Rules: Technological Change, International Competition and Regulation in Telecommunications*, Crandall, R., and K. Flamm (eds.), Washington, D.C.: The Brookings Institution, 1988.

[15] For a detailed analysis of this approach, see Thum, M., *Netzwerkeffekte, Standardisierung und staatlicher Regulierungsbedarf*, Tübingen: J. C. B. Mohl (Paul Siebeck) Verlag, 1995, pp. 150–157.

[16] David, P., and S. Greenstein, "The Economics of Compatibility Standards: An Introduction to Recent Research," *Economics of Innovation and New Technology*, Vol. 1, 1990, p. 29.

[17] See, for example, David, P. A., and M. Shurmer, "Formal Standard-Setting for Global Telecommunications and Information Services," *Telecommunications Policy* 20/10, special issue, 1996, pp. 789–815.

[18] The advantages of a committee approach over an interventionist one is discussed in detail in Woeckener, B., "Standardisierungspolitik für die Informationsgesellschaft," *Jahrbuch für Nationalökonomie und Statistik 215/3*, 1996, pp. 257–273.

[19] David, P. A., and M. Shurmer, "Formal Standard-Setting for Global Telecommunications and Information Services," *Telecommunications Policy* 20/10, special Issue, 1996, p. 803.

[20] David, P., and S. Greenstein, "The Economics of Compatibility Standards: An Introduction to Recent Research," *Economics of Innovation and New Technology*, Vol. 1, 1990, p. 25.

[21] Blankart, C. B., and G. Knieps, "State and Standards," *Public Choice 77*, 1993, pp. 39–52.

[22]   For an overview of public choice economics, see, for example, Cullis, J., and P. Jones, *Public Finance and Public Choice*, London, England: McGraw-Hill Publishing, 1993.

[23]   See, for example, Farrell, J., and G. Saloner, "Coordination Through Committees and Markets," *RAND Journal of Economics*, Vol. 19, No. 2, Summer 1988, pp. 235–252, or Weiss, M. B., and M. Sirbu, "Technological Choice in Voluntary Standards Committees: An Empirical Analysis," *Economics of Innovation and New Technologies*, No. 1, 1990.

PART

II

The standardization of advanced television services in the European Union

CHAPTER

4

Advanced television technology

4.1 Basic characteristics of a digital-TV system

The main emphasis of this book is on technology penetration and the accompanying process of standardization. Nevertheless, it is only reasonable that a discussion of the technical background be provided in order for the reader to have a greater appreciation of the economic and strategic issues. This is the concern of this chapter, which briefly explains the basic characteristics of analog, high-definition, and digital television [1].

The fundamental process through which a broadcast picture appears on a household TV screen can be broken down as follows:

‣ Production (filming) and processing of a program take place in a TV studio or on location.

‣ Pictures are transmitted via cable, satellite, or terrestrial networks (or a combination of these).

‣ A TV set in the household receives the signals (video and audio) and represents them in image form.

Between these three stages, it is the interfaces that are important. These interfaces require compatible technology. The broadcast chain is determined by two different standards: the studio or production standard and the transmission standard. The production standard determines how the picture is filmed and processed; the transmission standard determines how it is sent from the studio to the households. The household reception device, in turn, has to be adapted to the transmission standard.

In conventional TV systems, pictures are transmitted in analog form (i.e., in frequency waves). The standards for production, transmission, and reception are identical, but this is not a necessary condition. There are two reasons why it may be favorable to choose a different transmission standard.

First, transmission capacity on all distribution media is limited. Extending this capacity (e.g., launching an additional satellite or laying cable infrastructure) is very costly. For terrestrial transmission, the limits are set by national governments that allocate frequency spectrum to broadcasters. Due to scarce capacity, it is advantageous to be able to compress the studio signal so as to make better use of the available bandwidth. This is especially the case if the aim is to transmit high-quality pictures (i.e., high-resolution) that require even more bandwidth. It should be noted though that it is not desirable to use compression during production. This is because production requires multiple operations such as editing, recording, and effects, which may require repeated compression-decompression steps where the compression artifacts can accumulate and degrade the final picture quality.

Signal compression allows the transmission of more signals over a given bandwidth. Hence, it is possible to transmit more images in conventional quality or the same number of images but in higher quality. Video and audio compression requires technological specifications different from those of the production standard.

Furthermore, the signal is subject to external interference while being transmitted. To reduce the susceptibility of the signal to such interference, a transmission standard different from the production standard may be advantageous, so as to guarantee a picture quality that does not deteriorate during transmission.

Advanced TV systems, such as HDTV and digital TV address these issues. Digital TV technology, as illustrated in Figure 4.1, works as follows.

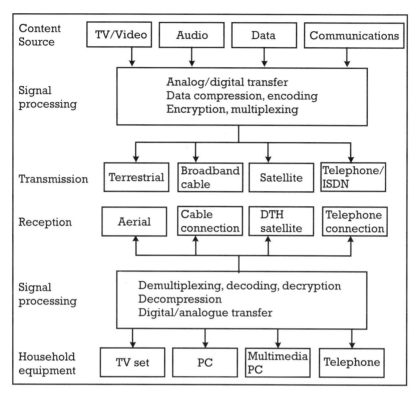

Figure 4.1 The components of digital broadcast transmission. (*From:* Prognos, "Digitales Fernsehen—Marktchancen und Ordnungspolitischer Regelungsbedarf," BLM Schriftenreihe Band 30, Munich, Germany: BLM, 1995, p. 10.)

The original (analog) signal is processed into a digital signal. This is achieved through a video compression standard, called MPEG-2, named after the international committee, the Motion Picture Expert Group, that developed it. MPEG-2 is used in digital-TV systems all over the world and can be regarded as a truly global standard [2].

Once compressed, the signal is transmitted digitally [i.e., in binary digits (0 and 1 bits)]. Digital technology is already widely used in other areas, particularly for computers. The process allows the transmission of up to 10 TV channels over the same bandwidth required for one analog channel. Furthermore, the picture quality can be altered according to the program requirements. For example, fast-moving sports programs are better to watch with a higher picture resolution, whereas low resolution is sufficient for studio programs such as talk shows. Finally, digital transmission is not prone to external interference. Hence, the picture does not lose quality while being transmitted.

At the receiving end (i.e., in the TV household) a digital signal can either be represented on a digital TV set, a so-called integrated digital TV (idTV), or it has to be reconverted into analog and can then be shown on an analog TV set. The conversion is done through a set-top box (also called a decoder box) that is inserted between the point of signal reception and the TV set.[1] A set-top box is very similar in size and connection mode to a video recorder.

4.2   Digital TV versus HDTV

In the 1980s, the focus of advanced TV developments was on high-definition images, which would provide better quality pictures for viewers.[2] HDTV involves the transmission of double as many image lines compared to conventional systems. A conventional color TV system has between 525 and 625 image lines and the image format is 4:3 (ratio of picture width to picture height). HDTV systems have a resolution of around double that of conventional systems and typically use 16:9 image format. This provides better picture quality, noticeable in particular on larger TV screens. At the same time, the wider screen format (similar to cinema screens) creates a more realistic image for the human eye. Consumers can, however, only benefit from improved picture quality if they have a specific HDTV set and if TV programs produced in high-definition format are available. To create HDTV programs, studios need new and costly equipment.

The HDTV standards available, high-definition multiplexed analog component (HD-MAC) (in Europe) and HiVision (in Japan), are hybrid analog/digital technologies. Video signals are transmitted in analog form but audio signals are digital. The signals are compressed in such a way so that an HDTV channel does not use more bandwidth than a standard-definition channel. The United States is the first country to have introduced an all-digital HDTV standard (see Chapter 10). In summary, the principal benefit of HDTV is the higher picture quality available to

---

1. The terms *decoder* and *set-top box* are used synonymously in this work.

2. The developments of HDTV technology and standardization are described in Section 4.6 (for Europe), Section 10.1.1 (for the United States), and Section 10.2.1 (for Japan).

consumers but this requires substantial investments at all stages of the process.

Digital television technology offers several benefits to consumers, and these result from two technological aspects. First, as digital technology is used in many other types of electronic equipment, content from sources other than broadcast production can be used. Equally, broadcast content can be received, for example, via the PC. All kinds of data such as sound, images, and text can be received via the TV set. This integration of various types of information (audio, video, text, and graphics) is defined as multimedia. Second, facilitated through signal compression, available transmission capacity is used much more efficiently. This allows the possibility of new consumer services such as the following.

> • *Multichannel TV:* Allows more channels to be transmitted in the same bandwidth, hence giving consumers a larger choice of programs available;

> • *Near video on demand (NVoD):* Allows a particular sequence of programs (that which generally constitutes a "TV channel") to be transmitted on several channels at staggered starting times (e.g., every 15 minutes) so that viewers can choose when to start watching a program;

> • *Enhancement of live events* (especially sports): Allows for the transmission of a sports, or other, event on several channels, with each showing a different camera angle, so that viewers can choose their preferred perspective;

> • *Complementary information:* Allows viewers, through the use of the remote control, to demand additional text or video information (e.g., when watching a football match, to get detailed information on single players);

> • *Interactivity:* Gives consumers, through a return (or feedback) channel, a direct influence on program choice or content (e.g., home shopping and video games), with the returned signals from the household to the service provider sent directly over the broadcast network or indirectly through a connected telephone line (enabled through a telephone modem included in the set-top box);

▸ *Data services:* Allows for the receipt of Internet or electronic data services via TV and may eventually also facilitate videoconferencing by TV [3];

▸ *PC reception:* Allows for the receipt of transmitted video, audio, and text signals via the TV set or a connected PC, enabling, for example, the downloading of software from the Internet;

▸ *Additional distribution media:* Uses data reduction to transmit certain types of content over distribution media with traditionally insufficient capacity, such as telephone (fixed and wireless) networks;

▸ *Video on demand (VoD):* Allows viewers to order at any given time a program from a "video library," which is then broadcast to them individually. In contrast to the above offers, the signals are not continually broadcast to all viewers (point-to-multipoint) but only on demand to a single household (point-to-point). VoD is available only over switched networks (i.e., networks that are capable of handling point-to-point communications). Such networks include telephone networks with broadband capacity (e.g., ADSL).

All these services can be received with a conventional TV set plus a set-top box. In addition to these benefits, digital TV also allows the transmission of higher quality images, including HDTV, enhanced-definition images, or wide-screen pictures (i.e., 16:9 rather than the conventional 4:3 format). To benefit from these features, however, consumers need to buy the necessary equipment (i.e., an HDTV or a wide-screen TV set).

4.3 Characteristics of different distribution media

As noted earlier, TV broadcasts can be received via satellite, cable, or terrestrial transmission. The capacities of these media differ. With regard to digital TV and the new services possible, the choice of distribution medium will be significant. Differences exist, in particular, with regard to transmission capacity, interactivity, geographical coverage, and household installation requirements. Furthermore, it has to be noted that existing (analog) channels are still in operation. Hence, for the next 10 years or so only a limited amount of transmission capacity can be used for digital broadcasts—that which is not already deployed for analog.

4.3.1 Satellite transmission

The main advantage of satellites is their high transmission capacity. Each satellite can transmit hundreds of channels. In Europe, around 350 analog TV and radio channels in more than 20 languages are available via satellite. This number can be (and is) further extended by launching more satellites [4]. Digital transmission will greatly enhance these numbers. Additionally, satellites cover large geographical areas. European satellite operators Société Européenne des Satellites (SES) and Eutelsat, for example, each operate several satellites that cover most of Europe. Satellite transmission is also widely used to feed TV channels into local cable networks.

To receive direct-to-home (DTH) satellite TV, consumers need to purchase and install a satellite dish. Installation can prove difficult if the design of the property prevents the satellite dish from being pointed toward the signal. In addition, local regulations may restrict installation on aesthetic grounds. In most countries, approximately 20% to 30% of households cannot receive satellite transmission due to either topography or planning restrictions or both [5].

To receive digital satellite TV (DST), viewers have to upgrade their satellite dish with an additional piece of equipment. This so-called low-noise block (LNB) currently costs around £30 to £50. Furthermore, current technology does not permit a direct return channel for interactivity. A telephone modem, included in the set-top box, enables a simplified return channel by telephone. Direct interactivity is expected to be possible within a few years [6].

4.3.2 Terrestrial transmission

Terrestrial transmission capacity is determined by national governments that allocate the frequency spectrum among broadcasters and mobile phone operators and for military and police utilization. Generally, all usable frequencies are allocated; hence, there is limited scope for capacity extension. In most European countries, terrestrial TV allows reception of around four to six analog channels.

Digital terrestrial TV (DTT) permits the use of additional frequencies. These so-called taboo frequencies cannot be employed for analog TV due to interference problems, but they can be used for digital broadcasting. In the United Kingdom, the first country with DTT services, capacity for around 36 digital channels has been made available.[3] This illustrates a crucial limitation in comparison to digital satellite and cable TV, which can offer hundreds of channels. Another important disadvantage of DTT is the lack of a direct return path. Any interactivity would have to be carried out (as for current digital satellite) via a telephone return channel.

DTT does offer considerable cost advantages. The consumer does not require a satellite dish or cable connection but merely a decoder box. Most TV households can receive DTT over existing roof- or set-top aerials. Furthermore, existing analog terrestrial infrastructure (i.e., transmission sites and masts) can be used. New digital antennas can be mounted on existing transmission masts, and digital transmitters can be located in existing transmitter buildings, thus reducing upgrading costs. A 1997 study by the Convergent Decision Group, comparing digital delivery system costs, found that DTT is the lowest cost delivery system, calculated per household covered [7].

Moreover, terrestrial TV has a high coverage. Around 95% to 99% of households in a country can receive terrestrial broadcast signals. This makes it ideal for fulfilling universal service obligations in the digital era, and it is a crucial distribution medium for portable TV sets operated with set-top aerials.

4.3.3 Cable transmission

As for satellites, one main advantage of cable transmission is the large channel capacity available. In addition to that, however, cable is the only distribution medium with real interactivity capabilities. Digital cable TV (DCT) can use a direct return channel that is of a higher bandwidth than the telephone line. This enables services such as "impulse pay-per-view" via the cable network rather than using the phone line. To the viewer, however, the difference is hardly noticeable. Unique to cable is VoD. The program choice of the viewer is decrypted in return for electronic payment using a smart card that is inserted into the set-top box.

Furthermore, the high interactive capacity enables video communications both to and from the home. Cable has another advantage in that it can combine TV and telephony. Cable operators can use available bandwidth to offer telephone services at lower costs than traditional telecommunications operators.

A significant disadvantage for cable is the high infrastructure cost involved in setting up a cable network with good household coverage.

---

3. The channel number that can be transmitted over the given capacity depends on the chosen image resolution. It is also expected to increase with further technology development.

The costs of "passing" a household vary principally with population density, hence urban areas are more attractive. To offer digital cable TV, existing networks also have to be upgraded. For European countries this is estimated at around £480 to £560 per TV household [8]. Once a digital cable network is in place, household reception equipment costs are lower than those for the other distribution media because no aerial or dish is needed, only the set-top box. Moreover, once a household is passed, connecting it for cable reception carries a minimal cost.

4.4 Additional technologies in the set-top box

In addition to the technologies required to convert an incoming digital signal into an analog one to be viewed on an analog TV set, the set-top box can contain a range of other components. The most important ones, conditional access, the electronic program guide, and the application programming interface, are briefly described in this section.

4.4.1 Conditional access

Digital TV is generally independent from its form of financing. The proliferation of channels implies that financing solely through advertising will not be possible [9]. The TV advertising market is largely saturated while at the same time costs for sports and film rights are increasing rapidly. It can therefore be expected that pay TV will become the dominant form of financing in digital TV. Pay TV, which already exists in analog, can mean the payment for a channel package (also known as a bouquet), for a single channel, a certain program (pay-per-view), or per viewing time.

The operation of digital pay-TV services requires a control mechanism through which only those viewers who have paid can watch a program. This is the process of conditional access (CA) [10]. A CA system consists of three subsystems: scrambling, encryption, and subscriber management. The first technology must be able to scramble and descramble transmission signals; the second technology has to control the descrambling process so that only selected receivers can view the transmitted signal. The way that scrambling and descrambling work at any instant will depend on a frequently changing number called the "control word." Whereas the scrambling process has no need for secrecy, the number must only be available to paying customers. Usually the number will be transmitted with the rest of the digital signals, but it has to be encrypted to keep it secret. At the receiving end, the control word has to be decrypted to permit the descrambling of transmitted content. If the encryption can easily be broken, the system is effectively worthless. The authorization for decryption is stored in a smart card, a card comparable to a credit card, which viewers insert into their set-top boxes. In addition to the CA components included in the set-top box, pay TV requires a subscriber management system (SMS). This is an administrative center that stores customer data and requests and issues smart cards and, eventually, invoices.

4.4.2 Electronic program guides

The multitude of TV channels and other services in the digital era may make it more difficult for viewers to find exactly what they want to watch. An electronic program guide (EPG), as perceived by the user, is a tool for browsing through information about the TV channels, programs, and any additional services available. Viewers can, for example, search for specific programs or for certain categories (e.g., comedy, documentary, and news) and can "bookmark" programs they want to watch or about which they want to be reminded. The EPG is frequently updated and can be customized to viewer preferences. In its simplest form, it is merely a digital version of the program listing one can find in daily newspapers. A more advanced version may combine such textual information with audiovisual information and provide a means to control different functions of the TV set or related products—for example, by allowing for automatic videocassette recorder (VCR) programming [11]. An EPG could also be compared to a portal site on the Internet in that it forms the first point of entry for viewers and has a great influence on the information they are likely to access.

4.4.3 The application programming interface

Interactive services require an application programming interface (API). An API is comparable to an operating system for a computer. It determines basic functions and a programming language. It is the basis on which different applications, such as home shopping, EPGs, or video games, can be developed and used. Applications can be downloaded to the set-top box. The API ensures that the downloaded software is correctly installed and functions without the viewer having to be involved.

Endnotes

[1] For a detailed analysis of digital TV technology, see, for example, Reimers, U. (ed.), *Digitale Fernsehtechnik—Datenkompression und Übertragung für DVB,* Berlin, Germany: Springer Verlag, 1995; *Digital Video Broadcasting—A Volume of Technical Papers Accompanying the Commission's Communication,* Brussels, Belgium: Commission of the European Communities, January 1994; de Bruin, R., and J. Smits, *Digital Video Broadcasting: Technology, Standards, and Regulations,* Norwood, MA: Artech House, 1999.

[2] On MPEG-2 technology, see, for example, Chiariglione, L., "Source Coding and Multiplexing." In *Digital Video Broadcasting—A Volume of Technical Papers Accompanying the Commission's Communication,* Brussels, Belgium: Commission of the European Communities, January 1994, pp. 15–46; de Bruin, R., and J. Smits, *Digital Video Broadcasting: Technology, Standards, and Regulations,* Norwood, MA: Artech House, 1999, pp. 140–155; and Challapali, K., and G. Nocture, "Video Compression for Digital TV Applications," *Philips Journal of Research,* Vol. 50, 1996, pp. 5–19.

[3] On technologies for Internet reception via TV, see Vision Consultancy Group, *Internet@digital.TV Report,* London, England, 1998.

[4] For an analysis of European satellite TV, see Ziemer, A. (ed.), *Digitales Fernsehen: Eine Neue Dimension der Medienvielfalt,* Heidelberg, Germany: R. v. Decker's Verlag, 1994, pp. 118–132.

[5] Convergent Decisions Group, *Digital Terrestrial Television in Europe,* London, England, Spring 1997, p. 8.

[6] For development trends in multimedia satellite communication, see Vision Consultancy Group, *Multimedia Satellites—The Ka-band Report,* London, England, 1997.

[7] For a detailed cost comparison between the different distribution media, see Convergent Decisions Group, *Digital Terrestrial Television in Europe,* London, England, Spring 1997, pp. 30–34.

[8] Convergent Decisions Group, *Digital Terrestrial Television in Europe,* London, Spring 1997, p. 33.

[9] Public broadcasting services, which are financed through license fees, face a specific situation resulting from the move to digital. This is discussed by Reiter, U., "Die Strategie der ARD im Digitalen Zeitalter," *Media Perspektiven,* No. 8, 1997, pp. 410–415, or Levy, D., *Europe's Digital Revolution,* London, England: Routledge, 1999.

[10] For more detail see Van Schooneveld, D., "Standardisation of Conditional Access Systems for Digital Pay Television," *Philips Journal of Research,* Vol. 50, 1996, pp. 217–225; Giachetti, J.-L., et al., "A Common Conditional Access Interface for Digital Video Broadcasting Decoders," *IEEE*

Transactions on Consumer Electronics, Vol. 41, No. 3, August 1995, pp. 836–841; and Reimers, U. (ed.), *Digitale Fernsehtechnik—Datenkompression und Übertragung für DVB*, Berlin, Germany: Springer-Verlag, 1995, pp. 171–176.

[11] On EPG technology, see Rosengren, J., "Electronic Programme Guides and Service Information," *Philips Journal of Research*, Vol. 50, 1996, pp. 253–265.

European standardization approaches for HDTV and DVB

5.1 The European Commission's policy on standardization

The adoption of digital TV implies a switch to a new system (or network) and requires considerable investment by the user. Viewers have to acquire a set-top box or a new digital TV set. Satellite viewers additionally need to upgrade their dishes. To justify this investment, users want to be certain that the new system will not become obsolete in a short period of time.

As noted in Section 2.2, systems markets are prone to "tipping," a phenomenon under which one system tends to pull away from its rivals in popularity once it has gained an initial edge. This tendency exists for TV systems [1]. To overcome user inertia or reluctance to switch to a new system, a single standardized TV network would be, *ceteris*

51

paribus, favorable to all players involved. This is a position generally agreed on by firms as well as regulatory bodies in Europe. A working paper published by the European Commission states the overall objective of broadcast standardization as follows:

> In broadcasting, the primary aim of standardization is to allow the inter-change of signals and programs. An ideal world would be one in which common standards were used everywhere, where programs produced in Japan, the United States, or Europe could be readily transmitted in any other part of the world, without the need for conversion, and where radio and television signals and receivers were common and inter-changeable. [2]

Opinion varies greatly, however, on how to achieve standardization. At present, there are three analog color TV systems being employed [phase alternation line (PAL) and Système Electronique Coul-eur Avec Mémoire (Secam) in Europe; National Television System Com-mittee (NTSC) in the United States and Japan)], which in themselves have variations in different countries. This is regarded as ineffi-cient. Common signal formats make it possible to achieve economies of scale in equipment production and thus bring unit costs down.

Standardization can take place at three levels: national, regional, or international. In the case of telecommunications, for example, inter-national standardization is pursued through the International Telecom-munications Union (ITU); the European body is ETSI. On the national level, standardization is carried out by the Ministry for Post and Telecom-munications (BMPT) in Germany and the British Standards Institute (BSI) in the United Kingdom; other countries have their own respective organizations.

In Europe, the principal trend is clear: Increasing privatization, deregulation, and product differentiation, together with an increase in cross-border operations, have resulted in regional standardization organi-zations considerably increasing their power. This has been achieved at the expense of national, and especially international, organizations [3]. In Europe, for high-tech markets, ETSI (for telecommunications) and CENELEC (for electronics) proved (relatively) swifter and more efficient in their results [4]. The trend toward regional standardization is also reflected in the race for competitiveness between Europe, Japan, and the United States.

In the late 1980s and early 1990s, however, the European Commission (EC) was not satisfied with the work of the European standardization organizations and pursued an active, interventionist policy. This was justified with the occurrence of a market failure in information and telecommunications technology markets. The EC's position is summarized in the following:

> At present, this transition to interactive trans-European networks and services is being held up by the fragmentation of markets, by insufficient interconnection and interoperability and by the absence of mechanisms to ensure coherent management.... Supply of services is inadequate and, where it does exist, too costly, with the result that demand is also too low as in this case it is supply, which determines demand. As a result, demand is not manifesting itself, which in turn discourages the creation of a viable supply. This is a vicious circle. [5]

In economic terms, the argument is that without full standardization a critical mass cannot be achieved because network effects cannot sufficiently be exploited. This, in turn, prevents new markets from emerging or developing.

It has been shown in a theoretical model that market failure due to a lack of standardization can arise if two incompatible systems are developed simultaneously [6]. When users perceive the two systems as "equal" (i.e., not differentiated) and when the market is divided equally among them, this can result in both not reaching critical mass (i.e., both systems will fail). For empirical evidence of such a situation, consider, for example, the introduction of two incompatible systems for quadraphonic sound in the 1970s, which resulted in a failure of both [7]. The EC justifies its intervention with the objective of preventing such a situation.

Since 1992, the legitimacy for EU intervention in standardization has been laid down in European legislation. According to Article 129c in the Maastricht Treaty:

> The Community shall implement any measures that may prove necessary to ensure the interoperability of the network, in particular in the field of technical standardization. [8]

With this provision, the EC is in principle able to survey and intervene in all standardization activities in Europe.

The theoretical discussion in Chapter 3 concludes that a direct regulatory (interventionist) approach is generally unfavorable. Sections 5.2 and
5.3 will analyze this conclusion on the basis of two empirical cases in the
audiovisual sector. First, Section 5.2 briefly outlines the European standardization process for HDTV. Here, the EC adopted a highly interventionist role. Subsequently, Section 5.3 will discuss the case of DVB in
detail. The latter forms a committee approach where the EC was only
"observing" the process.

It will be shown that, in accordance with the theoretical argumenta-
tion, the latter approach proved successful whereas the former failed.
Hence, the EC's decision not to make use of its power to intervene in
standardization has proven favorable for technological development in
the audiovisual sector.

5.2 HDTV

The European standardization process for HDTV forms an often quoted
example of a direct regulatory, interventionist approach that failed. Due
to the extensive literature already available on this case, this section will
not recall the developments in detail, but merely provide a summary of
the most important policy issues [9].

EU intervention in the HDTV standardization process was triggered
by several interrelated factors:

› Japan had developed an HDTV standard, called HiVision, and, in
  1986, proposed it as the world studio standard for HDTV to the
  Consultative Committee for International Radio (CCIR).[1]

› In the mid 1980s, both European industry and the public authorities
  were preoccupied with the challenges posed by the Japanese industry. Europe felt a strong need to get ahead of Japanese competition,
  especially in the high-tech industries, and felt obliged to protect its
  consumer electronics manufacturers. This view resulted particularly
  from the Japanese successes in video recorder technology (early
  1980s) and the VLSI semiconductor technology (mid-1980s). The
  position was strengthened by the fact that patents for the PAL and

---

1. See also Section 10.2.1.

SECAM technologies had expired, enabling non-European manufacturers to enter the market.

▶ TV set manufacturers were keen on launching a new TV system to revive saturated markets for TV sets. The industry traditionally had a strong lobby and demanded EU protection from non-European competition.

▶ Consistent with its focus on the single European market as the cornerstone of industrial policy, the EC aimed for common standards within the EU. For the TV sector this meant a migration away from PAL and SECAM toward the creation of a single TV standard.

In summary, efforts to create a genuinely European market for TV equipment and services without making it easier for Asian firms to dominate equipment markets had a major impact on strategies for HDTV.

European firms believed that the acceptance of the Japanese Hi-Vision standard would fatally weaken their competitive position in the one product area where they were able to retain market share against the Japanese, and hence lead to the eventual eradication of the European consumer electronics industry. The EU Commission's Task Force on IT (later DG XIII) and the EU Round Table of European IT Firms strongly supported the set maker's view and endorsed the argument that Europe should try to develop its own independent HDTV technology to compete with the Japanese.

As a result of these factors, Europe refused to accept the Japanese standard. The CCIR gave Europe four years to develop and propose its own standard. In 1986, led by the manufacturers Philips and Thomson and subsidized by the EC, the Eureka-95 (EU95) research project was launched. Under time pressure from the CCIR, the already existing multiplex analog component (MAC) technology was adopted and mandated for direct broadcast satellite transmission under the so-called MAC directive [10]. The focus was on the speed of development rather than on the pursuit of the "best" technology.

In April 1989, the Council of Ministers decided on a strategy for European HDTV that consisted of three closely linked areas of activity. Directives were passed on the setting up of norms for the transmission of television programs via satellite, on the support of technological development to enable European producers to provide the necessary equipment fulfilling these norms, and finally on the support of the audiovisual sector for efforts to supply services and produce programs with the new technology. These initiatives illustrate the strong involvement of the EC, first in the standardization process and subsequently in the technology implementation phase.

The 1986 MAC Directive failed to achieve its objective of securing a monopoly for the new technology in direct broadcasting by satellite, and European policy did not prevent the development of an alternative technology. In 1989, Sky Television started transmitting in the conventional PAL format through Astra, a low-powered telecommunications satellite. This type of satellite was not covered in the Directive, which was limited to high-powered direct broadcast satellite services. Similarly, in 1988, European public broadcasters, in collaboration with consumer electronics manufacturers, started work on a PAL-based system of wide-screen TV (PALplus) thereby also undermining the HD-MAC strategy of the EC.

Since the beginning of the 1990s, the European HDTV strategy faced increasing challenges from the rapid development of digital HDTV technologies in the United States. In 1993, the EC admitted that HD-MAC had become obsolete and that it would no longer support HDTV technology based on the HD-MAC standard. It was stated that the issue of standards was not yet decided and that the important issue was the introduction of wide-screen services, whatever technology used.

The highly centralized standardization approach adopted by the EC was driven by political interests rather than based on technological and economic reasoning. The actual standard selection was codetermined by business interests of the large set manufacturers (Philips and Thomson) and by political interests on the EU/government side. The EC saw its role as helping to establish a consensus between all actors in the audiovisual sector. In retrospect, it is clear that program makers, broadcasters, and, above all, consumers were left out of the crucial phase of standardization.

The HDTV case clearly illustrates that government authorities are unsuited for the job of setting technological standards. Industrial policy objectives and lobbying from influential companies distorted the process. Companies found a way to circumvent regulation, and technological development leapfrogged political decisions. Considering that at least ECU650 million were spent on the development and implementation of HD-MAC, this policy approach formed a very costly failure.

5.3 DVB

This section analyzes and assesses the emergence and workings of the Digital Video Broadcasting (DVB) group for digital-TV standardization in Europe.

5.3.1 The European Launching Group

With the realization that the hybrid HD-MAC standard was not successful and rendered obsolete with the evolution of digital technologies developed in the United States, by 1990 several members of the HDTV initiative had begun to perceive fully digital technology to be the way forward in television. By that time, industry players had moved away from a sole focus on providing higher picture quality, toward a focus on multichannel provision, with the choice of also offering improved picture quality. This shift was seen as particularly crucial to gain support from broadcasters. Broadcasters did not perceive any advantages from HDTV, but multichannel TV was seen as a significant potential revenue source. The further development of advanced television technology was based on the following objectives:

▸ A proliferation of the number of TV channels that can be transmitted over a transmission medium with a given (limited) bandwidth capacity;

▸ The transmission of audio programs in high quality, as well as data transmission for entertainment and business applications;

▸ A flexible choice of image and sound quality (including HDTV);

▸ For the use of pay services, the provision of secure encoding mechanisms, which prevent unauthorized access [11].

Initially, in Europe conversations took place between a few experts discussing the feasibility of current television technologies and the alternatives for the development of television in Europe. Peter Kahl from Germany's BMPT and Stephen Temple from the U.K. Department of Trade and Industry (DTI) initially agreed bilaterally to push the development of digital-TV technology. The intention was to provide a new starting point for the future development of TV technologies that would allow Europe to stay abreast of its competitors, namely the United States and Japan.

The EC and France initially opposed this new plan, especially in consideration of the strong support and large financial resources that they had provided for the HD-MAC initiative. Despite initial resistance, France, Scandinavia, and other countries soon recognized the attractiveness of digital transmission and joined the (then informal and confidential) European Launching Group (ELG) established in 1991. This brought together politicians of EU countries to assess opportunities for, and raise interest in, digital television technologies.

Satellite broadcasters were the first to show commercial interest in applying digital technologies because they enable the transmission of up to 10 channels on a single satellite transponder. Analog transmission only permits one channel per transponder. When it was clear that these broadcasters were prepared to invest money in these new technologies, the need to quickly determine common European standards was perceived so as to avoid a transmission standard fragmentation similar to that of PAL and SECAM.

5.3.2 The DVB Group

After the ELG had grown to comprise representatives from several European countries, it was decided in 1993 to launch it officially as the DVB Group. At that time the group comprised 80 members, including program producers and providers, network providers, and consumer-electronics and production-equipment manufacturers, as well as representatives from national and European regulatory bodies. Much of the initial momentum was derived from a group of senior public servants in the various European governments, who encouraged their respective nations to join. At an early point in the process it was decided that all interested parties should be involved, including non-Europeans.

The main objective laid down in a memorandum of understanding (MoU), signed in September 1993, was to determine transmission standards for data compression and the technological specifications for the different forms of transmission (mainly cable, satellite, and terrestrial). The agreed-on specifications would then be passed on to ETSI or CENELEC for official standardization.

The DVB project adopted an approach very different from those by official bodies or from the earlier HDTV initiative: It was able to include representatives of all relevant players of the audiovisual sector into a voluntary, dynamic, and consensus-based organization where decisions were to be led by commercial considerations.

As illustrated in Figure 5.1, the organization of DVB comprises two commercial modules (one for satellite and cable; the other for terrestrial transmission) with representatives of potential DVB providers and public authorities; a technical module open to technology experts from all signatories, and a steering board that observes the keeping of the MoU arrangements. The General Assembly, coming together once a year, oversees the project management and appoints members of the steering board. Working groups or ad hoc groups are formed for specific tasks.

Typically, the commercial module would formulate a set of user requirements to be adhered to by the specifications for a specific way of transmission. On the basis of this, the technical module would develop and propose technological specifications. In this way, it is ensured that the system was specified in accordance with the business demands of the services envisaged. After final approval of the steering board, the specifications would then be passed to the official standardization bodies for formal standardization.

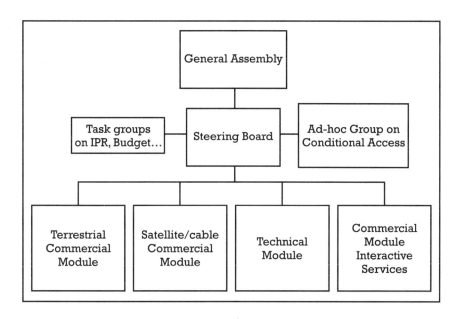

Figure 5.1 Organizational structure of the DVB group. (*Source:* Reimers, U. (ed.), *Digitale Fernsehtechnik—Datenkompression und Übertragung für DVB*, Berlin, Germany: Springer-Verlag, 1995, p. 7.)

This structure and development process was adopted with the objective of avoiding the mistakes of the HD-MAC experience where broadcasters' interests and market requirements had been ignored. The simple process described above proved highly efficient and generated a large number of specifications in a relatively short period of time.

In contrast to the lengthy processes of official standardization bodies, whereby interested parties give their views on a proposed standard, the DVB Group could quickly come to agreements on specifications since the interested parties were involved in the ongoing development process. Additionally, the speed at which transmission standards were developed could be achieved because the underlying technologies were usually not developed from scratch but existed already in some form or another and simply had to be "collected and combined" by the various technology experts involved. In this way, the DVB Group could avoid duplication of R&D work. Even final formal standardization by ETSI could be carried out quickly because ETSI members had already participated in the DVB development process.

Between 1993 and 1996, DVB agreed on the following family of standards [12]:

- MPEG-2: The MPEG-2 digital compression standard for audio and video transmission, developed by the international MPEG, was adopted as the basic transmission standard. MPEG-2 defines the way pictures are managed; it defines options for sound and provides the basic rules of the signal structure.

- DVB-S: Subsequently, the DVB-S specification for DTH satellite transmission was completed and standardized by ETSI in 1994. Although the initial technological interest in digital transmission was raised through the benefits for terrestrial transmission, especially the overcoming of spectrum scarcity, the DTH satellite operators (Astra and Eutelsat) and their satellite broadcasters first showed commercial interest and pushed the development of this specification. Moreover, the satellite specification was comparatively easy to develop, being based on already existing technology (the QPSK modulation format).

- DVB-C: Due to business interrelationships and competition issues, the cable broadcasters pressed for the cable specification to be developed shortly after the one for satellite. The cable specification, based on a 64-QAM modulation format, was standardized by ETSI in mid-1994.

▸ DVB-SI, DVB-TXT, DVB-CS: After having satisfied the immediate needs of broadcasters for satellite and cable transmission that enabled equipment manufacturers to get started on mass production, DVB turned its attention to additional issues—specifically the service information specification (DVB-SI)[2], the specification for the carriage of teletext in the digital multiplex (DVB-TXT), and a specification for satellite master antenna systems (SMATVs) (DVB-CS).[3] These were submitted to ETSI in autumn 1995.

▸ DVB-T: Terrestrial broadcasters initially did not show a great deal of interest in digital television; hence, work on the specification for terrestrial transmission was slow from the beginning. The terrestrial specification (DVB-T), based on the spread-spectrum orthogonal frequency division multiplex (OFDM) modulation technique similar to that used in Europe for digital audio broadcasting (DAB), was finally submitted to ETSI at the end of 1995 and approved in February 1997.

In the agreement and standardization of these specifications the industry consortia approach adopted by the DVB Group proved highly efficient and successful. However, the limits of a consensus-based, commercially driven standardization approach became obvious in the issue of CA. Section 5.3.3 discusses this issue.

5.3.3 The conditional access debate
As discussed in Section 3.3.3, at a high level of standardization the committee approach may encounter difficulties because firm-specific, vested interests are touched. This is precisely what happened in DVB with regard to CA. As explained in Section 4.4.1, two technologies are needed to control access: scrambling and encryption. The first process, requiring some form of scrambling algorithm, was quickly developed within the CA ad hoc group (chaired by Eamon Lalor, EC, DG XIII) and agreed on by all

---

2. Service information is used in DVB decoders as a basis for EPGs.

3. SMATVs are used to serve community antenna installations. The specification was adapted from DVB-C and DVB-S.

signatories. However, agreement on the second technology, CA, proved much more controversial. This is due to the commercial implications this technology has for broadcasters.

5.3.3.1 The commercial significance of conditional access

The technical standard of digital reception is regarded as the critical factor for gaining the best position in the new digital-TV market. For consumers, the ideal solution is an open system (i.e., a single technical reception infrastructure open to all content providers). This enables small broadcasters, such as special interest programmers or providers of multimedia services, to gain access to the market. Viewers would have free and individual choice of programs (or program packages) from all providers. Furthermore, economies of scale could be obtained in the production of decoders, thereby reducing their prices. Consumer electronics manufacturers also favor a common system as this enables them to mass-produce and sell the same equipment in a large market.

For broadcasters and/or service providers, however, different considerations apply. The central issue is the individual CA technology for the decoding of pay-TV contents. Whoever dominates the CA system and the SMS can achieve a very powerful position in the pay-TV business, being able to access subscriber data and, at the same time, control access of competing content providers. Network effects, as explained in Chapter 2, are likely to result in one dominant or only a few competing systems in each market. Naturally, each provider ideally wants its individual CA system to become the de facto industrial standard.

The same considerations apply for APIs and EPGs: Through control over an API, a broadcaster can control competing interactive applications or impose its own interactive applications onto other service providers. Providers of proprietary EPGs can control how viewers are guided through the channel offerings. Broadcasters with a disadvantageous placement on the EPG are likely to miss out on viewers. Hence, anticompetitive practices are possible if the program provider and EPG provider are vertically integrated. The dominant provider would list its channels first and those of its main competitors at the end of a channel listing.

That is why in the phase of market entry and market penetration, the set-top box, which contains these technologies, forms a critical factor. However, in the long run it is the contents (i.e., the ownership of rights for attractive programs) that determine market success. Only in the initial phase is the set-top box important due to the necessity to quickly gain large subscriber numbers to acquire an installed base. At the end of the day, it is irrelevant to the content provider which set-top box is being sold, as long as viewers subscribe to its contents. Providers of attractive content will gain the most significant revenues in the long run.

Therefore, CA and subscriber management, electronic program guides, and the API initially form bottleneck facilities. For example, an operator with a dominant position in CA systems has a "bottleneck" under its control. The operator assumes a gatekeeper position since the access of program providers to consumers depends on the permission of the CA provider. Such a monopoly position for CA can result in economic inefficiencies if market entry barriers are created for the market for TV programming. Control over CA is particularly problematic when the CA service provider is vertically integrated with a program provider. Vertical integration permits the gatekeeper to refuse access to the CA system or to provide access only under discriminatory conditions. It can be expected, for example, that the gatekeeper would prevent access of programs that directly compete with its own content.

The potential problems and the regulatory challenges arising with regard to CA APIs and EPGs are perceived differently in different countries. The CA debate that took place in the DVB Group is discussed in Sections 5.3.3.2 through 5.3.3.4.[4] The reactions of market players and regulators in France, Germany, and the United Kingdom are discussed in Chapters 6, 7, and 8, respectively.

5.3.3.2 The Simulcrypt proposal

The existing European pay-TV broadcasters (mainly Canal Plus, BSkyB, and Nethold) showed a very strong interest in developing proprietary, incompatible CA systems for their respective national markets and were not prepared to support a common European market for pay-TV services. Their position is based on two considerations:

▸ By keeping markets fragmented, piracy, which is already a big problem in analog pay TV, can be limited to the respective market segment (i.e., if one CA system is breached by pirates this will not affect the other markets).

▸ The now dominant pay broadcasters want to ensure that they keep their leading positions in the digital era and therefore intend to

---

4. APIs and EPGs were at this period not considered by the DVB Group. See also Section 5.3.5.

avoid European-wide competition. Furthermore, by operating a proprietary CA system, they want to make sure that they gain the revenues from undertaking the costly and risky move to digital.

In 1994, Canal Plus, News Datacom, BSkyB, and Filmnet proposed a solution on the CA issue to DVB called Simulcrypt. The main idea was that different broadcasters in different territories would use different CA systems. The various operators would partition Europe between themselves, taking responsibility for different zones. This would have the advantage of limiting the damage if pirates in a particular territory breached one system. Broadcasters that wanted to reach more than one territory would have to negotiate agreements with the operators of the respective territories to obtain access to their CA systems.

With the Simulcrypt system, the decoders in each country would be different so that, for example, Canal Plus smart cards would only work in Canal Plus decoders. The different operators would have special functions built into the interface between the CA subsystem and the decoders to keep their own security as high as possible. This was proposed as a complete and integrated European solution.

However, not everyone in DVB approved of this approach. In particular, public broadcasters and regulators—but also consumer electronics manufacturers—opposed proprietary systems claiming that they put too much power into the hands of a few powerful operators that would have control of the market and could, under pretexts, refuse access to potential competitors. To avoid such a situation, they proposed a different approach, Multicrypt.

5.3.3.3 The Multicrypt response
Simulcrypt opponents favored the so-called Multicrypt solution whereby a common interface (CI) integrated in the decoders would enable connection of different CA systems to a single type of decoder. This would allow a common European market for decoders as well as for programming material. The manufacturers of decoders (and eventually integrated receiver decoders) would be able to realize economies of scale in production. Moreover, transaction costs could be reduced and, most importantly, viewers would be able to choose from a larger variety of programming material.

As can be seen from Figure 5.2, the interface is between a "host" (the set-top box) and a detachable "module." The host contains an MPEG

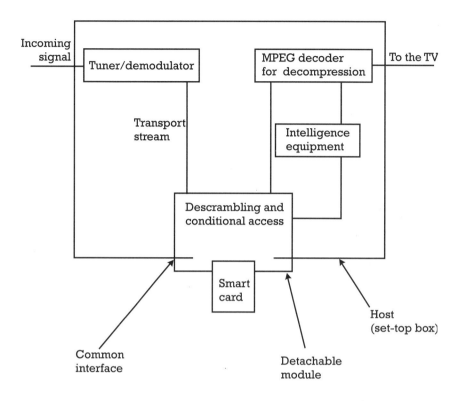

Figure 5.2 The CI. (*From:* Cutts, D., "Progress on Common Interface," *Advanced Television Markets,* No. 3, 1995 p. 7; Cave, M., and C. Cowie, "Regulating Conditional Access in European Pay-Broadcasting," *Working Papers Series No. 8,* Oxford, England: Regulatory Policy Research Centre, 1996, p. 18.)

decoder as well as a tuner and demodulator for cable, satellite, or terrestrial transmission, together with the intelligence that such a decoder needs. The CI lies between the standardized MPEG/DVB parts of the system and the proprietary elements for CA, which are here placed in an external module. It enables the separate delivery of these parts of the system.

The interface carries the whole of the demodulated MPEG data stream into the module. This is the same data that is delivered to the host over the transmission. The CA module then provides CA and descrambling functions and returns the stream to the host with selected services in descrambled form. The essence of the CI is that the host and module can interact without the host having to identify the particular CA system being employed.

The interface is generally extendible to allow new applications to be invented and delivered to hosts without requiring the host to understand them beforehand. It also allows upgrades of the CA and other functions to be implemented as they become necessary.

The economics of the Multicrypt option appear beneficial, especially when compared to a situation with a number of proprietary systems on the market. Its proponents emphasize that for the customer, acquiring a decoder with a CI will mean lower risk and lower cost. The equipment can be produced in higher volumes than if each model was designed specifically for one provider. This would allow economies of scale and encourage competition in the market for decoder supply. Moreover, the risk of consumers being "locked in" with a particular broadcaster can be avoided.

Although the use of a detachable module adds a small amount to the cost of the overall CA system, the volume increase available in the basic decoder production and distribution is estimated to more than compensate for this increment. Furthermore, the increased certainty available to both manufacturers and consumers is likely to speed up market adoption.

The argument was put forward that, since customers are not forced to purchase a particular type of decoder, the broadcaster can choose a CA solution freely or even run more than one CA system at a time. This would encourage competition among CA suppliers because the broadcaster is free to switch from one supplier to another. The possibility of upgrades to the system should also encourage technological development. The competitive environment is bound to encourage innovation and reduce cost. The Simulcrypt approach, in contrast, would create a fragmented market with smaller volumes of decoders controlled by major operators and the manufacturers that produce them. This environment of regional monopolies can be expected to raise costs for consumers at both service and hardware levels and denies them freedom of choice. It is claimed that by creating an installed base of proprietary decoder boxes to which they control access, the Simulcrypt providers effectively aim to lock out competition in the long run. Potential competitors would face enormous entry barriers. David Cutts, an expert involved in the DVB subgroup for the CI development, claims that it is "a measure of the importance of market control to these players that they continue to resist the general use of the interface despite the opportunity it gives them to lower their risk and costs" [13].

The proponents of Simulcrypt, in contrast, argue that only by employing proprietary technology, which guarantees future revenues, are they adequately rewarded for the risks they are taking in introducing new digital-TV services. Furthermore, the Simulcrypt proponents argue that pay-TV broadcasts require security from piracy [14].

5.3.3.4 The code of conduct on CA

With regard to CA, there was no hope of agreement within the DVB Group. Since it appeared impossible to reach a consensus, a voluntary, nonbinding code of conduct (CoC) was seen as the maximum agreement that could be reached given the strongly conflicting interests. The CoC would lay down the principles by which third parties would be granted access to proprietary CA systems. The provisions in the code intended to calm the fears of the non-pay-TV broadcasters concerning access to CA in a Simulcrypt environment. The basic objectives of the code are listed as follows:

- To facilitate an early CA infrastructure for digital TV services;

- To ensure that consumers will not need to acquire a multiplicity of decoders for receiving multiple digital-TV services from different providers.

Peter Kahl expressed his concerns over the length of time being taken to complete the CA package, especially with regard to the CoC. To speed up the process, it was determined that the entire CA package was to be put in front of the Steering Board on September 27, 1994, for final approval. If this were not possible, then a declaration would have to be made to the EC that a consensual agreement could not be reached. The consequences of this would be that the EC takes action and regulates matters related to the issue. All DVB members intended to avoid this option.

Concerns with regard to the CoC centered on the credibility it would have with those broadcasters that were the intended beneficiaries. Hence, the goal was to win a sizeable majority in favor of accepting the CoC so that it could be presented as part of the CA package to the DVB General Assembly.

To determine the opinions of the DVB members on the CoC, a questionnaire was circulated. The result was conclusive, with 80% of respondents in favor of accepting the voluntary and nonbinding code. It has to be noted, though, that only 63 out of 147 DVB members returned the questionnaire, hence the result is not really representative. In effect, this result means that DVB approves of both the Simulcrypt and the Multicrypt option. This principally leaves the decision as to the inclusion of a CI in the set-top box to the manufacturers. In practice, however, it is common in pay TV that the broadcasters assume control over set-top boxes. They order the boxes, including their licensed proprietary CA system, for production from manufacturers and distribute them to their subscribers. With the vote for the CoC, the DVB Group stated their belief that this arrangement would be sufficient to police the behavior of CA system providers.

5.3.4 EU regulation of digital-TV services
During the discussion on the CoC, the EC was already in the process of preparing a directive on the transmission of TV signals to replace the obsolete HD-MAC directive. The new piece of legislation was to be based on the work carried out by the DVB Group. Since the issue of CA had not been resolved, however, the directive was delayed and pressure was put on DVB to quickly resolve the issue, with the threat that otherwise the EU would take regulatory measures to guarantee fair and open competition (i.e., mandate a standard).

Meanwhile, the opponents of the Simulcrypt option strongly lobbied the European Parliament to support the standardization of CA technologies at a European level. The European Parliament recognized that the mandating of a CI could possibly delay the introduction of digital services by up to two years as it was expected that the CI would not complete the formal standardization process until late 1997. Simulcrypt, in contrast, was to be available by the beginning of 1996. Furthermore, European legislation on the key issues of CA and market control in the development of digital TV in Europe had to steer carefully between permitting the freedom needed for a digital market to emerge and the risks of generating incompatible standards across Europe controlled by a few powerful companies.

Faced with this dilemma, the European Parliament finally proposed a number of amendments to the draft directive. The most significant amendment with regard to CA is Amendment 5, with regard to Article 4 [15]. While recognizing that CA operators are entitled to earn a "fair return" on their investment, it proposes that broadcasters should be permitted access to CA services "on a fair, reasonable and nondiscriminatory basis" with access being policed by national governments. Furthermore, decisions on granting access are to become much more transparent, with providers of such services required to keep separate accounts for each level of their activities. The Directive 95/47/EC on the Use of Standards for the Transmission of TV Signals was passed with these amendments in October 1995 [16]. It mandates the use of the compression and transmission technologies agreed on by the DVB Group and standardized by ETSI. For the standardized specifications for transmission by satellite, cable, and terrestrial as well as those for modulation, channel coding, service information, and scrambling, the DVB members, under their MoU, have committed to make their own intellectual property rights available on terms that are fair, reasonable, and nondiscriminatory. In addition, a pooling arrangement for the common scrambling algorithm has been created, providing a one-stop shop licensing facility [17].

On CA, the directive states the following:

> Member States shall take all the necessary measures to ensure that the operators of conditional access services, irrespective of the means of transmission, who produce and market access services to digital television services offer to all broadcasters, on a fair, reasonable and non-discriminatory basis, technical services enabling the broadcasters' digitally-transmitted services to be received by viewers authorized by means of decoders administered by the service operators, and comply with Community competition law, in particular if a dominant position appears. [18]

Furthermore, pay-TV broadcasters that license their CA technology to manufacturers (as practiced already in analog pay TV) must not prevent the manufacturer from including a CI into the set-top box. Since the inclusion of a CI implies an extra cost to manufacturers, it is, however, unlikely that manufacturers would want to do this.[5]

For idTVs with a diagonal screen size larger than 42 cm, the Directive mandates the inclusion of at least one CI socket, permitting connection of

---

5. The extra cost of including a common interface is estimated at £40–50 for small production scales and £10 to £20 for large-scale production.

different CA systems or other peripherals [19]. This is significant because it implies that consumers can only be locked in during the initial market entry phase, when set-top boxes are deployed. Once idTVs become available and gradually replace analog TV sets, consumer lock-in through proprietary CA systems will no longer be possible. The fear of non-pay-TV broadcasters is, however, that it may be too late by then because the pay-TV broadcasters will already have firmly established their dominant position in the market.

The CA debate clearly illustrates three points. First, at a high level of standardization, where vested interests and preferences are strong, a consensual agreement is much harder to achieve than at the basic level. Second, there is a trade-off between compatibility (standardization) and the fast development of new markets. If pioneers such as BSkyB or Canal Plus are not allowed to gain a fair revenue on the costly and risky investment into the new digital market, they might hold back and thereby delay emerging markets. Finally, the CA issue has become quite problematic due to the vertical integration of the pay-TV broadcasters between the distribution level and the programming level. By controlling the CA technology, a vertically integrated company can exert monopoly power on the programming level, for example, by not granting access to the technology for potential competitors in content provision.

The debate around CA standardization was intense in the mid-1990s. However, with the introduction of digital-TV services, seeing competition evolving and lock-in effects not materializing, the issue gradually received less and less attention.

5.3.5 Further developments toward a common European API

As noted earlier, EPGs and APIs have not been considered in the initial DVB process. For EPGs there is no interest in a common standard. Broadcasters naturally wish to give their own brand, or look and feel, to their EPGs. Furthermore, EPGs can be customized by the viewer (similar to a portal site on the Internet); hence, the issue of potential competition distortion is disputable.

The debate surrounding APIs, in contrast, is similar to that over CA. As interactive services were initially seen as less important than multichannel pay TV in the development of digital-TV services, the discussion on API standardization was delayed until after the 1995 Directive. In January 1998, the DVB Steering Board announced that its members had agreed to adopt an "open" approach to interactive TV and that it would develop a common European API, to be called the multimedia home platform (MHP). In cooperation with the Digital Audiovisual Council (DAVIC), an international standards-setting body focusing on interactivity, the MHEG-5 standard was to be the way forward. This technology had the advantage of being a public standard and although the level of interactivity supported is low ("enhanced" teletext), many European broadcasters considered it to be sufficient for their initial digital-TV needs.

Canal Plus was quick to announce that it would incorporate the MHP specifications into its set-top boxes. Initially, this was seen as a success and as support for this technology. However, the fact that a large European pay broadcaster chose the standard looked to its opponents more like annexation than adoption. A group led by Télévision par Satéllite (TPS), Canal Plus' digital pay-TV rival in France and a supporter of the proprietary OpenTV API, opposed MHEG-5 as being inadequate to the task of dealing with fully fledged interactivity. Again, two sides evolved in the DVB Group, and no consensus could be reached. In July 1998, one month after its self-imposed deadline, DVB decided not to press for a solution but instead to concentrate on a next-generation API that would be based on the Java language. The specifications should have been finalized by mid-1999. However, industry experts estimate that the development of the technology will take until October 2001.

In the meantime, this consensus over the so-called Java-TV technology has come under attack from the United States. The Advanced Television Enhancement Forum (ATVEF), a U.S. organization headed by Microsoft and Intel, developed a specification for enhanced TV programming in the United States. This group advocates an approach to interactive TV, which is based on the hypertext markup language (HTML) and JavaScript, the two most commonly used languages on the Internet. They are incompatible with Java TV.

The choice of technology now depends on how fast an operable version will be presented. Furthermore, Sun and Javasoft, the developers of Java TV, have to agree to make their standard a public one. So far, it is a proprietary API that would not be adopted by the DVB Group. The specifications have to be licensed to any interested company at reasonable royalties. If royalties are set too high, the European API is unlikely to be widely adopted.

This illustrates an essential difference to the earlier standardization process: There is as yet no intention that Java TV is to be formally mandated by the EU. The long time it is taking to reach an agreement on a common API also illustrates that pressure from public authorities and the perspective of a mandated standard have a positive influence on rapid technology development. Despite the disagreement over CA, the out-come (in this case, nonstandardization), as laid down in the 1995 Direc-tive, was achieved relatively quickly. The final specification for a European API, the MHP, was formally standardized through ETSI only in 1999. Although the standard is not mandated by European law, several broadcasters announced they would employ MHP when they upgrade their set-top boxes. Commercial application is unlikely to start before October 2001. In the meantime, broadcasters continue to use proprietary technologies, such as OpenTV, Liberate, or MediaHighway.

It also has to be noted that the issue of a common European API is critical not so much for broadcasters and manufacturers, but in particular for European software developers that can gain significant economies of scale through a common system. The fact that they do not have such a strong lobbying power within DVB and in general in the EU may be another reason that the issue has not been resolved more quickly.

5.4 Assessment of the DVB standardization approach

Around five years after the 1995 EU Directive was passed, around 13 mil-lion European households received digital TV—and all through DVB technology. This implies a penetration rate of nearly 10% of households within four years from the first digital-TV service launch. This is a clear sign of the success of the DVB Group.

The DVB Group initially proved very fast and efficient at agreeing on standards. Agreement on basic compression and transmission stan-dards formed an important starting point. At these "low levels" of stan-dardization, network effects can be expected to be relatively high compared to the following levels of standardization, such as CA, APIs, and EPGs. It is these "low-level" standards that are essential to internalize network externalities and to benefit from economies of scale in decoder (and later idTV) production. The technology incorporated in the stan-dards has proven sufficiently superior to the existing analog technology. Furthermore, they are regarded as "future-proof": capable of handling developments such as new interactive services, VoD, or Internet-over-TV.

The combination of cooperation and competition ("coopetition") of European players within the DVB Group was achieved because all were interested in the deployment of common standards for Europe. In particular, the hardware manufacturers can gain significant economies of scale in the production of TV sets, set-top boxes, and related equipment. However, the prospects also proved attractive to broadcasters that had previously tended to limit their involvement to national markets. Standardization would enable the more rapid development of new services as well as easier access to new markets in other European countries.

At the initial stage, the advantage and efficiency of a committee process is clearly illustrated: In contrast to the interventionist HDTV standardization process, there was now no political pressure to develop a genuinely European standard. Technological reasons prevail; political influence can be avoided. Government subsidization is not necessary because the market players themselves are prepared to invest and carry the risks involved in developing this new market.

In the area of TV technology, the EC seems to have recognized that it is not suited to picking the winning technology. In fact, it is likely that the EC's restraint in regard to the DVB venture was a reflection on the previous failure in HDTV. The voluntary, industry-led DVB committee has efficiently agreed on standards for digital-TV services that have been adopted both in Europe and elsewhere.

It is also likely that the experience of the HDTV standardization process exerted wider influence. All players involved realized the mistakes of that approach, in particular the exclusion of broadcasters and the neglect of consumer preferences. This proved costly, especially for the EC, national authorities, and the large manufacturers. DVB emphasized the orientation toward commercial considerations and the inclusion of all involved players in the decision-making process. Furthermore, so as to raise international interest, the DVB Group invited participation from non-European players. This has resulted in the rapid adoption of DVB standards in many non-European countries, in particular in Japan (see Section 10.2).

Hence, the empirical analysis of DVB in this chapter reinforces the theoretical discussion on standardization approaches in Chapter 3: A committee approach to standardization is preferable to a direct regulatory approach. National authorities feared that a market-driven approach might result in a fragmented market or delay the introduction of digital-TV services. However, instead of intervening directly, which had proved inadequate in HDTV, the authorities chose to initiate a committee process. The outcome can be regarded as a great success. This success is generally acknowledged with regard to compression and transmission technologies.

There has been an intense discussion among industry experts, regulatory authorities, and consumer representatives on the fact that no common standard could be reached regarding CA. Manufacturers complained that they would have to produce different boxes for different broadcasters' services and non-pay-TV broadcasters claimed that they would become dependent on the dominant pay-TV operators. It has to be noted, however, that a large majority of the components required for decoder production are common to all "DVB-reliant" products. The CA system is just one component that differs depending on which broadcaster will use the boxes. Furthermore, due to the presence of network effects, it can be expected that in each market only a very limited number of CA systems will be deployed. Hence, any loss in supply-side economies of scale is likely to be small.

If a common CA system had been agreed on, it can be expected that the penetration of digital-TV services would have been much slower: Broadcasters would have been reluctant to provide the investments to persuade consumers to switch to digital, since there would be no possibility for them to recoup these costs in the future. This is because no lock-in effect is created. A fast market penetration is, however, also in the interest of manufacturers. With the prospect of producing and selling set-top boxes and idTVs in large numbers, the manufacturers have achieved their objectives. Advanced TV technology has already revived saturated TV equipment markets.

Another argument disqualifying the manufacturers' complaint of noncompatibility of CA systems is that boxes are incompatible anyway with regard to the medium of transmission (cable, satellite, terrestrial). This is due to technical limitations and not a result of noncoordination. The outcome, however, is similar: Some economies of scale are lost due to different components being necessary for different means of transmission. Hence, manufacturers require separate production processes. They can, however, design the boxes so that the changes required by different systems and CA modules will be segregated to one or a few small areas of the product. These can be modularized so that a single production line could produce different systems by substitution of these modules during production. This will minimize the disruption caused by having different systems. The additional losses from including different CA systems can be expected to be minimal.

In the future, boxes for different distribution media will be made compatible. This can evolve either through gateway technologies or by including the technological components for all types of distribution into each set-top box (or idTV) once production costs have come down.

With regard to the manufacturers' concerns, it can be concluded that the (small) losses in supply-side economies of scale that result from the present incompatibility of CA systems form the initial investment (or "sponsoring") necessary for manufacturers to create the new market.

The concerns of non-pay-TV broadcasters and regulators relate to demand-side economies of scale. They fear that once a pay-TV broadcaster has achieved an installed base through lock-in to its proprietary CA system, the market would tip in its favor and enable the broadcaster to establish a long-term dominant position—possibly a monopoly. The validity of this argument rests on lock-in. It has to be assessed, therefore, whether lock-in occurs and how strong and durable it is.

As digital pay-TV services form a new network market, it can be expected that incentives (in the form of set-top box subsidies) are needed to convince consumers to switch from conventional TV to digital pay TV. Subsidies lower switching costs, but at the same time reduce the lock-in effect. Moreover, if boxes are rented, as is common in analog pay TV, there is no significant lock-in effect at all. Furthermore, as noted earlier, set-top boxes are only transitory equipment. Eventually, idTVs will become the means to receive digital-TV services. Hence, lock-in is only temporary as idTVs have to be equipped with a CI. It can thus be concluded that lock-in, if at all, is likely to be weak and only temporary.

The present developments with regard to API standardization illustrate the benefits of the public authorities monitoring the process and exerting some pressure toward faster progress. When no consensus could be reached with regard to CA, nonstandardization, which is likely to be the favorable outcome, was quickly determined, and a formal mandate was achieved shortly afterwards. In the case of a European API, the EC did not state its intention to mandate a standard, and this lack of pressure is likely to be one reason for the delay.

A further argument in favor of the DVB outcome on CA is that the incompatibility of CA systems creates competition in the new market. This is likely to speed up development of this market and to create favorable conditions for consumers. This is precisely what all DVB members intended.

The arguments brought forward here are based on the theoretical findings discussed in Chapter 2. To further support the theory, Part III analyzes the actual developments in the market for digital pay TV.

Endnotes

[1]   This is shown in Farrell, J., and C. Shapiro, "Standard Setting in High-Definition Television," *Brookings Papers on Economic Activity: Microeconomics*, 1992, p. 177.

[2]   Wood, D., "Prospects for Common Worldwide Digital Television Systems." In *Digital Video Broadcasting—A Volume of Technical Papers Accompanying the Commission's Communication*, p. 289, Brussels, Belgium: Commission of the European Communities, January 1994.

[3]   See Drake, W. J., "The Transformation of International Telecommunications Standardisation." In *Telecommunications in Transition: Policies, Services and Technologies in the EC*, pp. 71–95, C. Steinfield, et al. (eds.), California: Sage Publications, 1994.

[4]   See Besen, S. M., and J. Farrell, "The Role of the ITU in Standardisation," *Telecommunications Policy*, Vol. 15, No. 4, 1991, pp. 311–321. For a detailed analysis of ETSI, see Besen, S. M., "The European Telecommunications Standards Institute," *Telecommunications Policy*, Vol. 14, 1990, pp. 521–530; for CENELEC, see Schellberg, K. U., *Technische Harmonisierung in der EG*, Europäische Hochschulschriften Reihe 8, Lang Verlag: Frankfurt/M., 1992.

[5]   Commission of the European Communities, "Growth, Competitiveness and Employment: The Challenges and Ways Forward into the 21st Century," White Paper, Luxembourg, 1994, p. 94.

[6]   Woeckener, B., "Standardisierungspolitik für die Informationsgesellschaft," *Jahrbuch für Nationalökonomie und Statistik*, Vol.215, No. 3, 1996, pp. 257–273.

[7]   See Postrel, S. R., "Competing Networks and Proprietary Standards: The Case of Quadraphonic Sound," *Journal of Industrial Economics*, Vol. 39, 1990, pp. 169–185.

[8]   Commission of the European Communities, *Treaty on European Union*, Office for Official Publications of the European Communities, Luxemburg, 1992, Art. 129c.

[9]   For a detailed description of the HDTV case, see, for example, Cawson, A., "High Definition Television in Europe," *Political Quarterly*, Vol. 66, No. 2, 1995, pp. 157–173; Cawson, A., X. Dai, and P. Holmes, "The Rise and Fall of HDTV: The Impact of European Technology Policy," *Journal of Common Market Studies*, Vol. 34, No. 2, 1996, pp. 149–166; Hart, J. A., and J. C. Thomas, "European Policies Toward HDTV," *Communications & Strategies*, Vol. 20, 1995, pp. 23–61; Heuser, U. J., *Hochauflösendes Fernsehen: Fallstudie und Analyse des Internationalen Standardisierungsprozesses*, Dissertation, Köln: Universität, 1992; Kaitatzi-Whitlock, S., "European HDTV Strategy: Muddling Through or Muddling Up?" *European Journal of Communications* Vol. 9, 1994, pp. 173–192; Peterson, J., "Towards a Common European

Industrial Policy? The Case of High Definition Television," *Government and Opposition 28/4,* 1993, pp. 496–511; and Watson Brown, A., "HDTV—A Personal View," paper based on a presentation to the DG XIII Technology Watch Group on April 1, 1993, Brussels, Belgium, 1993.

[10] Commission of the European Communities, "Council Directive (86/529/EEC) on the Adoption of Common Technical Specifications of the MAC/Packet Family of Standards for Direct Satellite Television Broadcasting," *Official Journal of the European Communities,* November 6, 1986.

[11] Reimers, U. (ed.), "Digitale Fernsehtechnik—Datenkompression und Übertragung für DVB," Berlin: Springer-Verlag, 1995, p. 11.

[12] For a detailed description of the technology, see Reimers, U. (ed.), *"Digitale Fernsehtechnik—Datenkompression und Übertragung für DVB,"* Berlin, Germany: Springer-Verlag, 1995.

[13] Cutts, D., "Progress on Common Interface," *Advanced Television Markets,* No. 3, 1995, p. 11.

[14] For a detailed discussion of the piracy issue, see Morgan, G., "The Conditional Access Dilemma," *Advanced Television Markets,* No. 26, 1994, pp. 14–15.

[15] European Parliament, *Amended Proposals for a European Parliament and Council Directive on the Use of Standards for the Transmission of Television Signals,* COM (94) 0455 (1994), Brussels, Belgium, November 11, 1994.

[16] Commission of the European Communities, "Directive 95/47/EC of 24 October 1995 on the Use of Standards for the Transmission of TV Signals," *Official Journal of the European Communities,* November 23, 1995.

[17] See DVB, "DVB Project Promotes Pooling of DVB Patents," *Digital Video Broadcasting,* press release of DVB Group of May 29, 1997.

[18] Commission of the European Communities, "Directive 95/47/EC of 24 October 1995 on the Use of Standards for the Transmission of TV Signals," *Official Journal of the European Communities,* November 23, 1995, Art. 4c.

[19] Commission of the European Communities, "Directive 95/47/EC of 24 October 1995 on the Use of Standards for the Transmission of TV Signals," *Official Journal of the European Communities,* November 23, 1995, Art. 3.

PART

III

Diffusion of DVB technology

CHAPTER

6

Contents

Digital TV in France: A European success story

6.1 Introduction

In France, digital broadcasting services were introduced as early as April 1996 and have been very successful. Within four years, nearly three million households subscribed to digital services. Moreover, against all predictions, the market has been able to sustain two competing digital satellite operators, Canal Satellite Numérique (CSN) and TPS. This chapter examines the conditions and strategies that have led to France's success and argues that the competition has actually contributed to the fast consumer adoption of digital TV.

Section 6.2 discusses the market structures in the French TV industry, analyzing why these structures are favorable to the launch of digital-TV services. Section 6.3 examines the business strategies of France's digital satellite broadcasters, while Section 6.4 analyzes the digital offerings of French cable operators. Subsequently, Section 6.5 looks at plans in France for DTT, and Section 6.6 gives an overview of the French regulatory

81

framework in the broadcast industry. Finally, Section 6.7 summarizes the developments in France.

6.2 The television landscape prior to the launch of digital offerings

The existing structures in a broadcasting market greatly influence the potential for the rollout of digital services. The French market displays several features conducive to the successful introduction of digital pay-TV services.

6.2.1 Terrestrial television

The large majority of French households (around 87%) receive a relatively small selection of terrestrial programs, consisting of public channels France 2 and France 3 (under the umbrella company France Télévision), La Cinquième and Arte (which share one channel), and private broadcasters TF1 and M6. Moreover, French terrestrial-TV offerings are subject to strict content regulation, particularly with regard to movies (see Section 6.5). In terms of audience share, the French terrestrial-TV market is almost equally distributed between private and public broadcasters (see Table 6.1).

Public broadcasters France 2, France 3, Arte, and La Cinquième together reach an audience share of 47%. TF1 with an audience share of 35% is the market leader in terrestrial television and gains around 50% of all TV advertising revenues. M6, a commercial channel focusing on a young audience, has continually been gaining audience share over the last few years.

Table 6.1
Analog Television in France: Technical Reach and Viewer Shares

	Technical Reach (%)	Viewer Share, December 1997 (%)	Viewer Share, December 1998 (%)
France 2	99	24.1	22.5
France 3	99	16.3	17.1
Arte/La Cinquième	89	1.8/1.9	3.0/4.2
TF1	99	34.7	35.3
M6	90	12.7	12.9
Canal+	77	4.7	4.6
Others	—	3.8	0.6

The almost 5% audience share for the terrestrial, analog channel Canal+[1] is relatively high considering that this is a pay-TV channel. Important determinants of its success are its comprehensive content rights for movies and premium sports (e.g., Premier League Soccer) as well as the fact that it reaches 77% of French households. The bias toward movies and sports has proved attractive to French viewers, despite Canal+'s being a relatively expensive way (F175 per month)[2] to receive just one additional channel. Canal+'s price is the highest of all the pay-TV offerings in Europe (see Table 6.2).

In summary, the relatively small terrestrial offering with strict content limitations forms a significant incentive to spend money in addition to the annual F700 license fee for the reception of more channels with attractive content.

6.2.2 Satellite TV

In comparison to other European countries, satellite television developed rather late in France. The first French direct satellite system, Télédiffusion de France (TDF), proved unsuccessful in the 1980s and only from 1992 did satellite TV begin to make an impact. At that time, CanalSatellite, a 70%-owned subsidiary of the Canal Plus Group, was launched. It broadcasts the Canal+ channel as well as the cable channels via France Télécom's Télécom 2A satellite.

Table 6.2
Pricing of Analog Pay-TV Packages in Europe (January 1998)

Company	Offering	Monthly Price (French francs)
BSkyB Basic	30+ basic channels	119
BSkyB Premium	30+ basic channels and 6 premium channels	299
Telepiu	2 premium channels	159*
Premiere	1 premium channel	160*
Canal+	1 premium channel	220*

*Includes decoder rental.
Source: Salomon Smith Barney (1998), Canal Plus Company Report 2/98, London, 1998, p. 13.

1. Note that "Canal Plus" refers to the name of the company, whereas "Canal+" refers to the company's premium pay-TV channel.

2. One French franc is equivalent to approximately $1.37.

The launch of CanalSatellite amounted to the introduction of a new distribution medium. The distribution strategy followed that of the cable operators: segmenting subscribers and with multichannel packages offering them greater flexibility in the range of services taken. In particular, the introduction of pay-per-view services for movies and sports events, which allows consumers with lower preparedness to pay for a premium service to make occasional use of such services, can raise additional revenues. In fact, although the Canal+ channel is experiencing a slowdown in subscriber take-up, the launch of satellite services (analog and digital) resulted in a strong overall increase in Canal Plus subscribers.

Satellite TV in France is attractive for two reasons. First, 70% of the population lives in areas without cable, and second, there are around three million Arab-speaking inhabitants who via satellite can receive public service channels from Morocco, Egypt, and Algeria.

Despite these favorable conditions, CanalSatellite had a slow start. Within five years, it gained a mere 1.2 million subscribers. The slow development of satellite, just as for cable TV, is also due to high subscription prices. Basic subscription (six channels) costs F135 and the premium package costs F330 per month.

In general it is clear that satellite penetration at only 5.7% out of 21 million households (1997 figures) is comparatively low in France. The respective figures, for example, are 16% in the United Kingdom and 12% in Germany [1].

6.2.3 The cable-TV market

In 1982, the Mitterrand government initiated a large-scale project, the "Plan Cable," with the objective of connecting 15 million households by 2000 with a complete fiber-to-the-home network. This network was intended not only for television but also for a range of interactive broadband multimedia services. The telecommunications companies were to finance the infrastructure, with the government supervising local operations of the services. However, as the costs rose to three times the original forecasts, successive governments gradually trimmed the plan to a mixture of fiber-optic-to-the-curb and coaxial lines to households.

6.2.3.1 Structure of the market

The industry is dominated by three companies. NC Numéricable [formerly Compagnie Générale de Videocommunications (CGV)] is majority-owned by Canal Plus. Until September 1997, when Canal Plus increased its stake to 76.6%, NC Numéricable was owned by the utilities company Compagnie Générale des Eaux. Noos (formerly Lyonnaise Cable)[3] is majority-owned by Lyonnaise des Eaux. The third cable operator, France Télécom Cable (FTC) is a subsidiary of France Télécom. These three companies hold about three-quarters of all cable subscriptions and 85% of passed homes (i.e., homes where cable is laid) that are connectable. Smaller, private operators such as Vidéopole, a subsidiary of the electricity company Electricité de France (EDF) or Mediareseaux, offer services similar to those of the large cable operators, but have much smaller networks.

Of the 2.3 million cable households (as of December 1997), more than 1.5 million subscribe to the basic service, costing F90 to F175, and comprising the terrestrial channels, 10 thematic channels, and some foreign language programs. The remaining households are merely connected to a cable network, and for a small fee (F30 per month), receive the so-called Service Antenne, which comprises the terrestrial channels and a few additional channels only.

6.2.3.2 Reasons for slow cable take-up

The cable industry in France has been characterized by a vicious circle of low revenues, large losses, insufficient funds for quality content and marketing, and poor subscriber growth, leading to low revenue generation. This unfavorable development can be attributed to several reasons.

- Government control of both infrastructure development and service operations slowed down work progress and dampened the private initiatives required to develop and market content.

- The development of Minitel under the Plan Télématique at the same time usurped some of the functionality potential of interactive cable—for example, on-line services, travel information and reservations, and home banking, services that were now available on the telephone-based Minitel.

- New terrestrial networks (La Cinquième, M6) started terrestrial broadcasting in 1986, thereby reducing the incentive for viewers to pay subscription charges for cable connection.

---

3. Lyonnaise Cable was renamed Noos in early 2000. For simplification, the new name Noos is used throughout the text.

▶ The launch of the terrestrial pay-TV channel Canal+ in 1984, offering an attractive combination of films and sports, proved a huge success and further reduced the potential for cable which had poor content in comparison. Moreover, high basic entry costs (F130–F180) for cable customers made cable services uncompetitive compared to the premium pay-channel Canal+ (F175). Only since the beginning of the 1990s has Canal+ been made available on cable networks.

▶ The ownership structure of the cable network is peculiar. France Télécom, which laid most of the network, owns a network share in more than half of the homes passed (6.6 million homes are passed) but only has 300,000 subscribers. The private cable operators Noos and NC Numéricable own part of the infrastructure but mostly use that of France Télécom. Paris Cable, for example, is controlled 52.4% by Noos, 24.5% by France Télécom, and 23.1% by the City of Paris. Unlike Deutsche Telecom in Germany, France Télécom even owns the connection from the curb into the household, the so-called last mile to the customer.

▶ The cable operators conclude long-term (20–30 years) franchising contracts with the local authorities, thereby gaining a geographical monopoly. The networks built under the Plan Cable, laid and owned by France Télécom, are rented out to private operators. The rental fee to be paid depends not on the size of the network but on the number of subscribers. This type of pricing means that the cable operators are unable to realize economies of scale from increasing subscriber numbers. Moreover, the cable operators finance the thematic channels, which are set up to increase the attractiveness of cable TV. Per month and subscriber, the cable operators pay around F5 for each thematic channel in their bouquet. Due to this peculiar pricing structure, cable subscription is relatively expensive. For example, Noos' Service Antenne costs between F15 and F36, the basic package F120 to F135, and two movie channels an additional F75 to F90 per month. More importantly, this pricing structure means that the cable operators have no real incentive to gain more subscribers, and of course, this is reflected in low penetration rates (passed households in relation to connected households) for the large cable companies: NC Numéricable 18%, France Télécom Cable 24%, and Noos 19% (1998 figures).

In contrast, small private cable operators such as Vidéopole that built their own networks as a response to the failure of the Plan Cable offer lower prices. The basic service here costs less than F90 per month. As a result, penetration rates, in the range of 26% to 35%, are more favorable. However, the small private operators taken together only account for about one-fourth of total cable subscriptions.

It is only since the beginning of the 1990s that cable has shown significant growth. New channels and some interactive services (e.g., home shopping) have been introduced, and growth rates of around 13% per annum have been observed. At the end of September 1998, 2.5 million households were connected to cable, whereas 7.1 million homes were passed (i.e., 35% of connectable households make use of the cable option—for detailed and the most recent figures, see Section 6.4). By comparison, in Germany, the figure is almost double, at 56%, whereas in the United Kingdom it is even lower, at 23%.

Total cable penetration (cable subscribers in relation to TV households) as of November 1998 still was relatively low at 12.5%. However, it is as yet unclear what killer application will pull in more cable subscribers. Nevertheless, it is worth mentioning that most cable networks are ready for digital transmission and that all cable operators have started to offer digital bouquets. The fierce competition between CSN and TPS can even prove favorable to cable to develop its customer base: Canal Plus charges F6 per month for the carriage of its thematic channels whereas TPS only demands F2 to F3. Such price competition might help to drive down costs for cable operators and, thereby, consumer prices for cable connection.

6.2.3.3 Cable-TV regulation

In addition to the peculiar structures within the cable industry, the industry is also subject to strict regulation. Cable is fighting a difficult battle against satellite TV, which, due to its cross-border nature, enjoys a deregulated environment in France, being able to undermine content restrictions and offering ever more channels, many of them foreign. Unlike satellite, cable comes under terrestrial-TV legislation, which provides constraints on offerings (e.g., quotas for French and European content and channels for ethnic minorities). This also means that satellite channels, when fed into a cable network, have to respect these regulations. This forms an important constraint on the cable operators' ability to develop attractive packages and gain subscribers in the weak cable market. The possibility of digital transmission, which allows distribution of around eight times as many channels on the available capacity (30–40 analog channels), has encouraged cable operators to press for a liberalization of their program regulation, to be able to compete with satellite.

So far, the government has only half-heartedly attempted to introduce some measures favoring cable over satellite. For example, a 1993 law determines that all new houses in cable franchise areas have to be passed by cable. Furthermore, local authorities can prohibit the installation of satellite dishes for technological or aesthetic reasons. That is why in central Paris, for example, there are hardly any satellite dishes to be found.

6.2.3.4   The French television market has growth potential

This analysis of the French television landscape provides some clues to the potential for future development. Due to the particular structure of the French TV market, with most viewers still only receiving a relatively limited terrestrial offering, there exists a considerable incentive to pay extra for the reception of premium movies and sports, but also for a larger variety of programs (i.e., basic multichannel packages). Furthermore, it appears that existing pay-TV offerings such as Canal+ or the thematic cable and satellite channels are too expensive to draw in significant subscriber numbers. The real potential, therefore, is in a price-competitive, multichannel offering. Canal+ has so far benefited from its monopoly position, which allowed the company to charge monopoly prices. Under competitive conditions, prices will have to be reduced.

6.3   The digital satellite broadcasters

France saw the launch of three digital satellite services, CSN, TPS, and ABSat, between April and December 1996. This section looks at the players involved and their competitive strategies.

6.3.1   Canal Plus: The first mover in the market

Canal Plus, the only established pay-TV broadcaster in France, was first to launch a digital service, CSN. Section 6.3.1.1 illustrates the company's strong position in the European broadcasting market. Section 6.3.1.2 then looks at the CSN offering.

6.3.1.1   Company background

Canal Plus is the largest and most vertically integrated European broadcaster. With around nine million pay-TV subscribers in June 1997, the

Canal Plus Group has clearly been the leading operator for analog pay TV in Europe (see Table 6.3). To understand the significance of this company's impact on the European pay-TV market, a brief overview of the company's activities is given in this section.

Canal Plus' operations in France   The core of Canal Plus' business remains the premium channel Canal+ launched in 1984 in France. Despite pressures on profits caused by the introduction of digital TV and by increased programming costs, the service is highly profitable and provides the group with a steady cash flow. Canal+ generates 76% of total group revenues and, in 1996, yielded a net profit of F1.3 billion.

By the end of 1997, Canal+ had 4.3 million individual subscribers or nearly 20% of French households (cable, satellite, and terrestrial). The channel has now reached maturity, with annual growth slowing down to around 3%, as the bulk of viewers prepared to pay high rates for a premium channel have already signed up. Additionally, the launch of Canal Plus' multichannel services—CanalSatellite in 1992 and CanalSatellite Numérique in 1996 (see Section 6.1.3.2)—might have an impact on subscriber growth although the premium channel Canal+ is available to CanalSatellite subscribers at a reduced price (F155 instead of F175). To extend its subscriber base, in September 1997 Canal Plus increased its 20% stake in one of the largest cable operators, CGV (now NC Numéricable), to a controlling stake of 76.6% (see Section 6.4.1).

Canal Plus' international television operations   Canal Plus focuses on the French market but also has international ambitions in other European countries. Language- and country-specific variations of the French Canal+

Table 6.3
Percentages of European Pay-TV Subscribers (June 1997)

Company	Subscribers (%)
Canal Plus	43
BSkyB	34
Premiere/DF1	8
Modern Times Group	1
Others	14
Total	100

Source: Salomon Smith Barney, Canal Plus Company Report 2/98, London, 1998.

channel are broadcast in Spain, Belgium, Italy, The Netherlands, and Poland.

In April 1997, the group acquired the pay-TV operator Nethold at a price of £1.24 billion, giving Canal Plus access to most of Europe's pay-TV market outside Germany and the United Kingdom and around 1.6 million additional subscribers in Europe, thereby increasing its total subscriber base by 20%. Shortly after the merger with Nethold, Canal Plus swapped its 37.5% stake in Premiere (Germany) with Kirch's 45% share (plus F9 million in cash) in Telepiù (Italy). As a result, Canal Plus now controls 90% of Telepiù.

In August 1997, an announcement was made stating the consolidation and expansion of the group's European presence especially in Scandinavia, Italy, Belgium, and The Netherlands. Among the measures being implemented is a plan to make greater use of the group's experience in France and eventually to unify the foreign subsidies' individual strategies.

Following the acquisition of Nethold, Canal Plus relaunched all of the Nethold operations, orienting them more closely with the French offerings. With low subscriber penetration (below 7% in each country), high churn rates[4] (e.g., 40% in Scandinavia) and a costly digital launch in Italy, Nethold was and continues to be a loss-making enterprise. In total, Canal Plus counts around 883,000 subscribers in Scandinavia. A digital platform, Canal Digital, was launched in May 1998 and gained 358,000 subscribers by September 2000. In Belgium, FilmNet was relaunched as Canal Plus Flanders, offering a three-channel package at Bfr 1,350 per month to a subscriber base of 165,000.

Canal Plus' Spanish interests are conducted through Sogecable, in which the group has a 25% stake, the maximum ownership by any entity permitted under Spanish regulations. The digital offer Canal Satellite Digital, launched in early 1997, quickly gained 260,000 subscribers but was then challenged by a competing service, the government-backed Via Digital, which followed in October 1997. By September 2000, Canal Plus had a total of 2.7 million subscribers in Spain—960,000 of those digital. In Poland, the Canal+ premium channel has 347,000 subscribers, and the new digital platform Cyfra+ gained 270,000 customers.

Beyond Europe, CanalsPlus has interests in North Africa and in Japan. In Africa, the group established pay-TV services through its 80.7%

---

4. The churn rate determines the number of subscribers who do not renew their subscription.

ownership of Canal+ Horizons. This service broadcasts both from Intelsat and Eutelsat and by September 2000 had attracted 133,000 subscribers from Senegal, Tunisia, Morocco, and the Ivory Coast.

In Japan, Canal Plus has interests through its subsidiary Multithématiques. This company, a multinational group founded by Canal Plus, General d'Images and Tele-Communications International, creates and distributes thematic channels around the world. Multithématiques signed an accord with Japan's Jupiter Programming to create two channels for cable and satellite distribution in Japan.

More importantly, Canal Plus has been looking for a large international partner. The objective is to form a strong, multinational opposition to media giants, such as Bill Gates of Microsoft, who is now entering broadcasting markets. Finally, in February 1998, Warner Bros., owned by Time Warner, purchased 10% of CanalSatellite. This deal, estimated at F400 million, represents the first major investment by any U.S. film studio in a European TV venture.

Thematic channels and film production   While Canal Plus is primarily a pay-TV operation, revenues generated from other activities have become increasingly significant, most importantly the provision of thematic channels and film production.

Thematic channels, carried on both cable and satellite, generate revenues both via carriage fees and advertising, but are still loss-making. In the same way as basic cable programs are provided in the United States, the French thematic channels are financed by (low) fees from cable and satellite operators as well as through advertising. Therefore, despite only small market segments, losses are kept low. Moreover, as shown in Table 6.4, most thematic channels are controlled by Canal Plus, and contents can be used at several different times.

Canal Plus' film production and distribution activities are conducted through a number of subsidiaries such as Le Studio Canal+ (100% ownership), Canal+ Video (100%), UGC-DA (97.6%), Canal+ DA (75%), and Ellipse Programme (69.3%). The group has been aggressively expanding its presence in the production and distribution business, mainly through acquisition, thereby increasing the proportion of the group's revenues contributed by production and distribution to 12% of total group revenues in 1999.

A well-positioned player to move to digital TV   Despite the acquisition of Nethold and the increased stake in NC Numéricable, both loss-making,

Table 6.4
Canal Plus' Thematic Channels (1999)

	Program Type	Ownership (%)	Total Subscribers (1,000s)
Eurosport	Sports	33.0	3,630
i-television	News	100.0	1,538
Canal J	Children	18.1	2,600
MCM	Music	20.0	3,116
Planète	Documentary	30.5	2,713
Forum Planète	Documentary	30.5	1,365
Canal Jimmy	Classic series	30.5	2,290
Paris Première	Entertainment	15.0	2,707
Ciné-Cinemas	Movies	30.5	729
Ciné-Classics	Classic movies	30.5	631
Monte Carlo TMC	Entertainment	23.7	3,002
Seasons	Hobbies	30.2	106
Muzzik	Music	19.9	533
C:	Software download and information	80.0	7
Demain!	Employment offers	99.8	1,374
Comedie!	Comedy	18.7	1,684

Source: Canal Plus, *Annual Report 1999*, p. 52.

Canal Plus promises stable future subscriber and earnings growth as a leader in European pay TV. Although the group now faces competition in some of its major markets, namely France, Spain, and Scandinavia, the risks to Canal Plus from this competition are limited: Subscribers and profits are growing continuously, the group has secured key content in exclusive long-term contracts, and, most importantly, Canal Plus' track record in pay TV is probably one of the best in the world, with fast subscriber growth, break-even achieved soon after launch, and very low churn rates (8% to 9% in France). Due to the group's diverse geographical exposure and strong cash generation of its premium service in France in particular, Canal Plus will be able to cover losses in other business areas for some time.

The extensive experience and financial resources of Canal Plus made it the best-prepared player to enter the digital pay-TV market. The Canal Plus Group launched its terrestrial pay-TV service in 1984 and has since been active in channel operation, service distribution, content provision, and technology management. Canal Plus has been following a successful commercial strategy incorporating vertical integration, diversification, and internationalization. Most importantly, it holds a large range of content rights, especially for U.S. movies and premium sports. For example, an exclusive five-year output deal was signed with 20th Century Fox in 1997. As the only pay-TV operator in France, pioneering digital broadcasting in France was intended to safeguard its monopoly position in the future as well.

6.3.1.2 CSN

CSN, which was launched on April 27, 1996, is commercially operated by CanalSatellite.

Reception technology   Its initial program offering was similar to the analog CanalSatellite, but the digital version is now distributed via the Astra 1E satellite. This means that existing satellite subscribers wishing to change to the digital offering have to upgrade their satellite systems to make it Astra 1E–compatible. Additionally, an explicit decoder rental of F45 per month has to be paid, whereas the CanalSatellite (analog) decoder rental is included in the subscription price. Because of the company's intention to terminate the analog satellite service, new customers could only subscribe to the digital offering. By the end of 1998, all customers had been converted to digital.

CSN is using its own decoder technology developed by SECA, including the proprietary MediaGuard CA system and a technology allowing interactive services, called Mediahighway. The provision of rental decoders at a relatively low price to the viewer means a high capital expenditure for Canal Plus. The company, which has to buy the decoders in bulk from manufacturers to keep the price low, has to depreciate these costs over time. This, in effect, implies a heavy subsidization of decoders. For example, in 1996, for its French services, Canal Plus charged F535 million as depreciation for decoder subsidies to its profit and loss account.

The CSN program offering   To provide attractive content, CanalSatellite has purchased rights from five of the seven Hollywood majors (i.e., the large U.S. film studios). Moreover, in 1996, it acquired UGC-DA, France's largest film library with around 5,000 films, and most importantly, Canal Plus purchased the exclusive rights for Premier League soccer, Formula 1, and rugby. The television rights (especially movies and sports) form a major competitive advantage for the company.

As of 2000, the basic digital multichannel offer, CanalSatellite Thématique, consists of around 50 television and several radio channels at F99 monthly. Premium options include, for example, Classique and Jazz (F27), Disney Channel (F35), Cinéma (five movie channels at F52), Découverte (six documentary and info channels at F79). The pay-per-view service Kiosque provides movies and sports events at F25 to F45 per event. Movies cost between F25 and F35, dramas or concerts F45, Premier League Matches F45, and the Premier League Season F640. For an additional F155, Canal+ can be received on four channels at staggered starting times as well as in a wide-screen version.

Interactive services on CSN   In combination with a PC, software such as video games can be downloaded through the decoder. This option, called C:Direct, costs an additional F50, excluding the purchasing price for software. C:Direct produces TV programs about games as well as an optional download service via satellite providing subscribers with a large assortment of shareware and demos as well as full-specification software at a discount of around 20% off retail prices. Payment is by bankcard using the Canal-Satellite decoder. C:Direct is thought to form the hub of Canal Plus' planned high-speed Internet by satellite service. By 1998, C:Direct had attracted 70,000 subscribers, but there was no further growth in 1999.

Further interactive services offered by CSN include the following:

> ‣ An electronic program guide;

> ‣ CanalSat Jeux (video games);

> ‣ Demain! (employment offerings);

> ‣ PMU direct (sports betting via TV);

> ‣ La Chaîne Météo (interactive weather service);

> ‣ Eurosport (interactive information complementing the programs);

> ‣ CanalSat Finance (financial services);

> ‣ CanalSat Boutique and Club Téléachat (interactive shopping).

With the introduction of second-generation set-top boxes in early 2001, CanalSatellite plans to offer full Internet access via TV.

Early digital success   The challenge for Canal Plus is to convert existing subscribers to Canal+ into subscribers to both Canal+ in digital form and CSN, as well as to introduce new consumers to multichannel TV in the hope that they will subsequently upgrade to the Canal+ premium channel. The risk is that existing subscribers to Canal+ will churn away from the premium channel and leave customers satisfied with the cheaper CSN multichannel package alone. So far, Canal Plus appears to have achieved its goal, with 58% of digital subscribers taking both CSN and Canal+. This allowed the company to increase the average revenue per subscriber by as much as 70%. For example, in 1998, monthly revenue from these dual subscribers was estimated at F300 per subscriber [2].

The competition from TPS, requiring significant additional marketing expenses, as well as heavy foreign expansion and the development of digital had resulted in losses of around F700 million in 1997 and F480 million in 1998 for Canal Plus. Still, with revenues and subscribers continually climbing, the French digital service reached break-even in July 1999.

Within only 18 months, Canal Plus' digital offering gained a subscriber base of 700,000. However, 60% of the new digital subscribers were then existing analog customers (i.e., to a certain extent, the digital service has been "cannibalizing" the existing service). During 1998, CSN subscriber numbers grew by 43% to 1.12 million, with a much larger share of new customers. Since then, however, growth has slowed. By the end of 1999, CSN counted 1.4 million subscribers (24% annual growth), but this only grew to 1.5 million until October 2000.

It can be stated, thus, that Canal Plus, despite facing competition on its "home turf" (i.e., the French pay-TV market) for the first time has managed very swiftly a transition to digital and could maintain its dominant position in the pay-TV market.

6.3.2 TPS: Challenge to the dominant player

The launch of TPS in December 1996 was greeted with much skepticism among industry experts. It was generally agreed that each national market in Europe could only sustain one single digital provider—and not two incompatible platforms.

6.3.2.1 The emergence of TPS

Initially, the largest commercial broadcaster in France, TF1, intended to cooperate with Canal Plus in setting up a digital satellite offering; however, Canal Plus did not show any interest. As a result, TF1 gathered other partners and created a serious competitive threat to Canal Plus instead.

TPS is a cooperative project set up by the terrestrial free-TV operators in France to provide a counterweight to Canal Plus. The shareholder structure illustrates this. TPS is owned by TF1 (25%), M6 (20%), CLT-Ufa (20%), France Télévision (8%), as well as Lyonnaise des Eaux (10%) and France Télécom (17%). In February 1998, CLT-Ufa sold its stake, worth F395 million, to M6 (5%) and Lyonnaise des Eaux (15%), raising their stakes to 25% each. However, CLT-Ufa still maintains an indirect stake through M6, of which it holds 40%. With all national free-to-air broadcasters and two cable operators participating in the venture, the TPS bouquet, in contrast to CSN, was not positioned as a premium offering. Instead, it focused on provision of the existing terrestrial channels in digital quality via satellite.

The free-TV operators decided to set up TPS as a competitor to Canal Plus for three main reasons. First, the French market is perceived a suitable market for digital satellite pay TV: Cable has been rather unsuccessful, and there are only five to six terrestrial channels. Second, Canal Plus holds a monopoly in pay TV and is seen as overcharging customers. This implies that there is room for competition at lower prices. Finally, the advertising market in France is saturated. Advertising growth, which used to be at 7% to 8%, now is at only 1% to 2% per annum. Therefore, broadcasters have to either squeeze their profit margins or expand the market. The move to digital pay TV is considered a useful expansion of the market. Moreover, with a penetration rate of 27% in 1997, pay TV has for a long time been firmly established in France; hence, it can be expected that the threshold for viewers to pay for television services is lower than in other European countries. For example, penetration rates for pay TV are at 3% in Germany, at 13% in Spain, at 9% in The Netherlands, and at 23% in Britain.

6.3.2.2 Reception technology

The TPS program offering is distributed via the Eutelsat Hotbird satellite. To receive the bouquets, a digital decoder with France Télécom's Viaccess CA system is required. The decoder is also equipped with the OpenTV API, an operating system that allows the use of interactive services. The decoders are rented at F45 per month or can be purchased at F3,000.

6.3.2.3 The TPS program offering

TPS initially offered a basic package with 13 thematic channels at F90 per month. The package TPS Cinéma provides three movie channels with 150 films per month at F100. Both packages combined, called Tout TPS, are available at a cost of F130 per month. TPS also offers bouquets targeted specifically at immigrants. Arabesque, a bouquet of five Arab channels, is offered at F64 and the SIC channel (F15) targets Portuguese-speaking households. Furthermore, subscription to TPS includes the six national channels; a selection of radio stations; the pay-per-view service Multivision, which offers, for example, Champion League matches; and several interactive offerings (see Section 6.3.2.5).

As of 2000, the TPS channel offer had grown to a total of around 130 TV channels, 45 radio channels, and 35 interactive services. They can be subscribed to in a range of combinations and offers. Remarkable in the TPS offer is that viewers can subscribe to the cinema bouquet without having to buy the basic thematic package. Although the films on this channel are not as up-to-date and as high-quality as those of Canal Plus, this option is still popular with viewers, as there are few movies in free-to-air TV.

6.3.2.4 TPS' competitive advantages

The reason for TPS' success as a second operator of digital pay-TV services can be accredited to several factors. Most importantly, the TPS bouquets are offered at a lower price than Canal Plus' channels. A price comparison is shown in Table 6.5. Although the price difference is only minimal for the basic thematic bouquet, the discount on premium channels is considerable.

The premium channels (Cinestars 1 and Cinestars 2) offer mainly movies in a market where relatively few films are broadcast on free-to-air

Table 6.5
TPS' Promotional Pricing Structure at Launch (Prices per Month)

Package	Pricing (French francs)	CSN Equivalent (French francs)	Discount (%)
Thematic	90	98	8.2
Premium Film Channel[1]	100	158	36.7
Thematic + Premium Film Channels[2]	130	253	48.6

[1] Compared to CanalSatellite + Film Options.

[2] Compared to Canal Plus Digital + CanalSatellite Basic + Film Options.

Source: Morgan Stanley Dean Witter, "The Usual Suspects: Potential for Digital Broadcasting in Europe," Memorandum 26.11.1997.

TV. The fact that the majority of those films are broadcast in the "second window" (i.e., have been shown once already on Canal+) does not appear to be a major deterrent. The price advantage here appears to be a decisive factor.

Furthermore, TPS has several promotional assets, such as the big TV companies TF1, M6, and France Télévision (which operates France 2 and France 3), which combine to achieve an audience share of 87.8%. On these channels, TPS is heavily advertised. The initial marketing campaign, based loosely on "the best thing since the invention of television" theme, is considered a small masterpiece among experts. Around F150 million have been invested for this, which quickly gained awareness and a reputation for TPS.

Another major advantage for TPS has been the fact that all free-to-air channels are included in the TPS bouquet. In particular, the public channels France 2 and France 3, as well as the commercial channels TF1 and M6, were exclusively broadcast on TPS' digital satellite offer until October 2000, when this exclusivity deal was ended through a change in the broadcasting law. Initially, La Cinquième and Arte were also exclusive to TPS; however, they were later made available on CSN. These channels, especially the public broadcasters, can terrestrially only be received in bad quality—or not at all—in many of France's mountainous areas. Therefore, satellite services see potential subscribers, who, for example, want to watch regional programs carried on the public channels. Hence, this provides viewers with an incentive to subscribe to digital satellite and means—at least a temporary—competitive advantage for TPS over CSN.

An important disadvantage for TPS is Canal Plus' policy of blocking access to pay-per-view and second-window pay-TV rights to French films. Canal Plus, investing heavily in domestic film production, discourages producers from selling those rights to competitors. As a result, TPS' pay-per-view offering (Multivision) had virtually no French films. This situation was improved in September 1997 when TPS signed a deal with Gaumont for certain pay-per-view rights and secondary transmission pay-TV rights (after Canal Plus).

6.3.2.5 Interactive services

An important aspect of TPS has been its strong focus on interactive services. TPS offers a range of services including the following:

▸ An electronic program guide;

> L'Oeil du Hibou—a 24-hour interactive arts and culture information service;

> TPS Boutiques—an interactive shopping mall;

> Bandiagara—interactive video games;

> F1—financial services;

> Météo Express—interactive weather service;

> Info Express—news, cinema, sports information;

> Espace Annonces—classified ads.

Since June 2000, TPS has also provided e-mail and Internet access via the TV; however, customers can only access selected sites including discount travel site lastminute.com, banking and trading site ConSors.com, and video game site Gameplay.com.

TPS claims that 91% of its subscribers use at least one interactive service regularly. The company reported 450,000 commercial or bank transactions via TPS in 1999. The most popular service is Météo Express, which is used by 83% of its subscribers.

In the short run, TPS regards interactive services as an investment in its image, rather than as being especially profitable. Only in the long run, when the market can be extended from the niche for technology pioneers to mass markets can interactivity be expected to yield profits. It has to be noted also that CSN has responded quickly to TPS' interactive offers and, as noted in Section 6.3.1.2, by now supplies several similar services. Hence, the differentiation effect may be eroded.

6.3.2.6   A strong competitor to CSN

TPS has proved highly successful. Its number of subscribers by the end of 1997—350,000—was more than double of what the company had expected. Moreover, that figure had grown to over 600,000 by the end of 1998 and to more than 900,000 by mid-2000. These growth rates—approximately 60% a year—are much higher than those of CSN, and if they continue, TPS may well catch up with CSN in the near future.

6.3.3   ABSat: From platform operator to program supplier

ABSat started as a third digital broadcaster in December 1996. This venture, a new entrant to the broadcasting sector, emerged from AB

Productions, a producer of series and TV films for the low end of the market. Its main asset is a large content library. For example, a major client for AB Productions' content is TF1.

ABSat started broadcasting via a Eutelsat satellite but from September 1997 also transmitted its "AB Découverte" package of seven channels via the Astra satellite system. The decoder provided to subscribers is able to switch between Eutelsat and Astra. Initially, ABSat did not want to invest in decoders, and so they had to be purchased by viewers. This was later considered a deterrent to subscribers; thus, since November 1997, the company has rented decoders at F40 per month, only marginally cheaper than CSN or TPS decoders.

The basic multichannel offer "AB Numérique" comprised eight thematic channels and an electronic program guide for F49 per month. Two premium packages, "Cinéma Numérique," with five movie channels and "Passions Numérique" with six thematic channels, were available at F119 each (a price including the basic offering). The three bouquets together, "Tout ABSat," cost F199.

ABSat uses the France Télécom decoder equipped with the Viaccess CA system. In June 1997, a Simulcrypt agreement was signed with CanalSatellite allowing CSN subscribers to receive ABSat services without changing decoder or smart card. This, however, did not work in reverse. To enable ABSat viewers to watch CSN programming, France Télécom, which provides the Viaccess technology, would have to reveal technological specifications to competitor CanalSatellite. Only in December 1997 did France Télécom and CanalSatellite sign a licensing contract for the Viaccess technology, enabling ABSat and CanalSatellite to agree a Simulcrypt accord for the reception of both bouquets on each other's decoders.

For a long time, no similar agreement existed between ABSat and TPS, although both broadcast on Eutelsat satellites, and both use the same CA system, Viaccess. Only in July 1998 was it agreed that ABSat's six wholly owned thematic channels would be carried on TPS. This considerably extended ABSat's reach.

ABSat, however, failed to win a significant number of subscribers on its own platform and thus decided to end its operations as a platform operator. Instead, it developed into a supplier of thematic channel packages for the other two platforms and the cable operators. In addition, ABSat provides single thematic channels for broadcast as part of basic packages. By March 1999, 250,000 households had subscribed to an

ABSat subscription option via satellite and 34,000 through cable services. A year earlier, the respective figures were 65,000 and 12,000.

This development suggests that ABSat's offer is too small and too specific to sustain a separate network. On the other hand, it illustrated that independent program (or package) suppliers with a differentiated niche offering can succeed in enlarging the offers of the main pay-TV operators. The advantage of ABSat as an independent program supplier is that it can offer its content on several platforms and therefore reach a large number of viewers. The disadvantage is that it will only be accepted on the platforms if it is not in direct competition with the channels owned by the platform operators.

6.4 Digital-cable offerings

NC Numéricable, Noos (formerly Lyonnaise Cable), and France Télécom Cable, the three largest cable operators in France, are the main suppliers not only for analog but also for digital services. Sections 6.4.1 to 6.4.3 describe their strategies in detail. Section 6.4.4 then analyzes the digital cable situation in aggregate to assess its potential as a competitor to the digital satellite broadcasters.

6.4.1 NC Numéricable

NC Numéricable (formerly CGV) was created in 1984 and operates 31 cable networks (i.e., franchise areas with a monopoly for cable-TV supply). The company launched a digital package in December 1996 and within one year gained a subscriber base of 35,000, bringing total subscribers to 376,000. In 1996, the cable operator made losses of £47 million, but it is expected to break even by the end of 2001. This became more likely when Canal Plus, in September 1997, took control over the cable operator by increasing its stake from 20% to 76.6%. By mid-2000, NC Numéricable had 665,000 subscribers—15% of those, digital.

The acquisition of NC Numéricable further expands the presence of Canal Plus within pay TV in France and ensures that cable is not wholly dominated by companies associated with TPS, since Noos and France Télécom Cable (via its parent France Télécom) are both shareholders in TPS.

For Canal Plus, a link between cable and satellite delivery is perceived as complementary. Cable is regarded as ideal for densely populated city centers whereas satellite is best suited for less densely populated towns and rural areas. Moreover, in the highly competitive digital environment, a cable partner is seen as an advantage. Without the burden of large infrastructure investments, Canal Plus has taken several measures to reshape the company, lowering the entry cost to the service and increasing the build-out of the network. A new company strategy was implemented in February 1998, when the company name was changed to NC Numéricable, a new management was introduced, and an enriched program offering developed. In August 2000, in response to France Télécom Cable's move to more customizable program offerings, NC Numéricable also changed its subscription options.

The basic offering "Découverte" at F138 per month (including decoder rental) comprises the six national free-to-air channels, 20 thematic channels, nine foreign channels, and access to Canal Plus' pay-per-view service "Kiosque." On top of that, subscribers can choose from 37 additional channels: 3 at F45, 6 at F70, 9 at F90, and 12 at F110 per month. Other single channel options include the Disney Channel (F35), Seasons (F30), TPS Cinema (three movie channels at F70), and Ciné Cinemas (four movie channels at F55).

Alternatively, customers can subscribe to the prepackaged bouquet Evasion at F179 per month. It is aimed at the family and comprises the six free-to-air channels and 17 thematic channels.

NC Numéricable carries channels from Canal Plus as well as from TPS. Furthermore, the channel packages are different from those of CSN. This can be seen as a differentiation strategy, offering new customers alternatives depending on the distribution medium.

A distinct advantage of cable over satellite is the combination of television and telephony services. With the aim of developing a telephony and Internet access offering, in September 1998, Canal Plus agreed to sell a stake of 37% in NC Numéricable to Exante, a consortium of U.S. investors. The new telephone venture is to directly compete with France Télécom, the incumbent telecommunications operator. By March 1999, however, only around 5,000 customers used the Internet service, and telephony had not yet been offered.

6.4.2 Noos

Noos (formerly Lyonnaise Cable) is majority-owned by the utilities company Lyonnaise des Eaux, with the U.K. cable operator NTL holding 27% since July 2000. Noos operates 25 cable networks. In mid-2000, a total of

772,000 households subscribed to the cable-TV service—32% of those receive digital services.

Noos launched the digital service in June 1997, six months after its rival NC Numéricable and just ahead of France Télécom Cable. The launch had been delayed by the establishment of TPS in which both Noos and France Télécom Cable had invested. As with TPS, both cable operators use decoders with the Viaccess system, developed by France Télécom and supplied by Sagem.

Noos has developed even more personalized channel options where people can choose channel by channel from a range of 135 channels and interactive services. Subscribers pay a minimum of F110 per month, which includes decoder rental, the six free-to-air channels, five thematic channels, and access to the pay-per-view service Multivision.

All other channels are assigned a number of "stars" with more stars meaning a higher premium value. For example, the travel channel Voyage counts five stars whereas the (typically more expensive) Disney Channel is designated by 40 stars. Viewers select as many channels as they want and pay according to the added number of stars: 25 stars at F30, 75 at F80, 150 at F130, 225 at F170, and 300 at F210. The Infinity option comprises all channels at F360 per month. The channel options can be changed on a monthly basis. Subscription to the Canal+ channel is possible, but it has to be done separately with Canal Plus.

Additionally, Noos is rolling out high-speed Internet access via TV to its subscribers. The Noostvnet service is offered at the special price of F300 per month.

6.4.3 France Télécom Cable

France Télécom Cable, founded in January 1983 to commercialize the networks constructed under the Plan Cable, operates 140 cable networks with a total of 730,000 subscribers, of which 10% are digital.

France Télécom Cable launched its digital cable service in September 1997. By 1998, this was offered on nearly all of its networks. It had, however, only around 38,000 subscribers. This may be due to the fact that only a single basic package was offered at F150 per month. This was later split into two at F75 each.

In 1998, the operator was first to announce the intention to move toward greater choice and more flexible packaging—a result of detailed research on viewers' preferences. In September 1998, the operator introduced "Modulo Cable" in selected networks. To access this digital offering, the viewer has to pay F80. This provides viewers with the free-to-air national channels, the pay-per-view service Multivision, the employment information channel Demain, France Courses (horse racing), 2 shopping channels, 6 foreign channels as well as a digital decoder. In addition to this basic offering, the viewer can choose between six predetermined packages (Modulo Passions from F70) with 3 to 8 channels each, or assemble their own mini-package (Modulo Cartes) consisting of 3 (F60), 6 (F85), 9 (F105), or 12 (F120) channels from a group of 32. Premium offerings comprise the movie option from TPS and CSN. Three further single premium channels at F35 each are also available, and the digital version of Canal+ can be purchased (F155 per month), with a minimum subscription period of six months only. Notable in this channel offering is that viewers can subscribe to premium channels without having to buy a thematic package; however, they still always face the F80 digital network access charge. A further novelty is that subscribers can change their package choice every month (although the minimum subscription generally remains one year). With this strategy, the company aims to gradually introduce new customers to multichannel TV and to differentiate its products from the big basic offerings of other pay-TV operators. The company claims that this formula has drawn in twice as many subscribers as expected. Subscribers generally choose a low-priced à la carte package first and upgrade soon afterward to a more expensive option.

A trial for Internet access that had accompanied the introduction of Modulo Cable has been abandoned due to lack of customer interest. By March 1999, the service had gained only 5,000 customers. Subscription to this service had cost F285 per month, a price that included the modem.

6.4.4 Aggregate analysis of digital cable TV

By mid-2000, France had a total of 424,000 digital cable subscribers (see Table 6.6), which is a rather modest figure in comparison to the satellite broadcasters. Digital accounts for 15% of all cable-TV subscribers, but the percentages differ considerably among the cable operators. With 32% of all its subscribers receiving digital, Noos is the most successful in transferring its customers to the new technology.

With the introduction of digital, the cable operators have a new opportunity to compete with satellite in that they can additionally offer telephony and Internet services, which are not (yet) available over satellite. However, the cable operators have so far been rather slow in making use of their competitive advantage. By mid-2000, cable

Table 6.6
Digital Cable TV Subscribers in France, August 2000

Cable Operator	Total Cable Subscribers	Digital-Cable Subscribers	Percentage of Total Cable Subscribers
NC Numéricable	665,000	101,000	15.2
Noos	772,000	247,000	32.0
France Télécom Cable	737,000	76,000	10.4
Others*	773,000	8,000	1.0
Total	2,940,000	432,000	14.7

*Comprises around 18 smaller operators.

penetration (cable subscribers versus TV households) was still low at around 11%. Less than 15% of all cable subscribers received digital services; only 81,000 had subscribed to high-speed cable Internet; and only 30,000 used cable for telephony.

A barrier to fast penetration of digital TV and Internet via cable is the need for France Télécom, which owns the networks operated by Noos and NC Numéricable, to upgrade the network for digital use. In April 1997, the two dependent operators filed a complaint with the Autorité de Régulation des Télécommunications (ART), the French telecommunications regulator, against France Télécom, claiming that the latter would only upgrade the network at an excessively high charge to the operators and with a delay of two years. The ART decided in favor of the two cable operators and blocked the 150% charge increase.

The large investments needed to upgrade the network and this ART decision prompted France Télécom to announce in February 1999 that it would sell those cable networks that it does not operate itself. The Suez-Lyonnaise Group, parent company of Noos, has expressed its interest in purchasing the networks it operates. This would give Noos a greater degree of freedom to become a multiservice operator, as the U.K. cable companies have been for years.

In comparison to the satellite operators, digital cable has been slow to develop in France. This is due to three factors. First, the peculiar ownership structure of cable networks has inhibited the upgrade to digital. France Télécom, owner of most networks, has been reluctant to provide the investments needed, as it would be thereby supporting its competitors. Second, the regulator's delay in permitting the supply of multiple services via cable has, until recently, prevented cable operators from making use of their distribution medium's distinct comparative advantage. Both these barriers to cable operators' competitiveness result from government policies.

A further reason, which is applicable to both analog and digital cable, is the poor customer service. A 1999 study by the market researcher IPSOS found that 70% of cable subscribers are unhappy with the programs carried and find the services too expensive for what is offered.

Finally, the fact that the cable operators are all linked to one of the digital satellite broadcasters means that they have not been forced to compete and have thus been slow to move to digital. Digital cable offerings, at present, constitute a different distribution medium with different package options and possibly some additional services, but they do not imply separate platforms. It remains to be seen if this will remain so in the long run. Whereas NC Numéricable is largely owned by Canal Plus, the other two cable operators are still separate companies that are linked to TPS through their parent companies' stakes in the satellite service. They may get into the awkward situation of having to compete with TPS for subscribers. It is more likely, though, that the companies will try to complement each other and concentrate on the competition with Canal Plus. An indication for the latter is France Télécom Cable's removal of two Canal Plus channels (Planète and Canal J) from its network in 1999. They were replaced with similar channels from TPS (Odyssee and Teletoon).

6.5 DTT: Still to come to France

The final distribution medium to be considered is DTT. There has been little interest from broadcasters in DTT; however, the government has launched several studies on the viability of DTT.

From a technological point of view, France appears a rather unsuitable market for DTT. Terrestrial broadcasting requires a large number of transmitters to achieve national coverage. This is due to the mountainous terrain and the existence of six terrestrial networks, which take up available frequencies. The costs of launching a DTT service to more than 60% to 70% of the population are estimated to be prohibitively high [3].

Furthermore, the existing players have not shown an interest in developing another means of distribution. Cable and satellite operators regard it as another competitor; Canal Plus opposes it because it would cannibalize its existing terrestrial package, which still is successful; and the free-to-air broadcasters are not interested because they believe that their domain would be diluted. Moreover, most of the free-to-air broadcasters are involved in TPS and would therefore face cannibalization of this service through DTT.

A white paper, commissioned by the French Ministry of Culture and published in April 1999, evaluates economic and legal aspects of DTT, including demand and possible new uses for this transmission medium. The report recommends DTT launch for 2001 or early 2002. It suggests awarding six multiplexes, each carrying four to six channels. Four of those multiplexes, with coverage of 60% to 80% of the population, are to be used to simulcast existing free-to-air channels. The investments needed for this project are estimated at F3 billion. Unlike in the United Kingdom, the French plan for DTT emphasizes the opportunity for local and regional broadcasts. The report envisages analog switch-off to take place between 2010 and 2015.

In late 2000, the government started the process for allocating the digital multiplexes, and broadcasters are preparing for a program offering that may rise from the current 6 channels to 36 channels. DTT broadcasts are to commence in mid-2002.

6.6   The regulation of broadcasting in France

Regulation can have a significant impact on digital-TV development and vice versa. This section lays out the regulatory structures in France and their effect on digital-TV penetration, as well as the impact of digital technology on the regulatory framework.

6.6.1   The regulatory authorities

Responsibility for broadcasting regulation in France is taken by the Conseil Supérieur de l'Audiovisuel (CSA). It is an independent media authority operating on the basis of the 1986 broadcasting act. The CSA consists of nine commissioners and a number of services for certain tasks with around 200 people involved.

The areas of responsibility of the CSA include the following.

  ▸ Licensing of terrestrial and cable operators and interactive services;

‣ Ensuring pluralism (especially in news);

‣ Youth protection;

‣ Media competition;

‣ Monitoring of advertising and sponsoring;

‣ Frequency planning and coordination;

‣ Technical supply optimization;

‣ Advice to the government on media issues.

The CSA has the de facto priority on issues of content regulation. In addition, the CSA is responsible for the monitoring of public broadcasters.

Telecommunications regulation is carried out by the ART. So far, the responsibilities of the two regulatory authorities have been kept distinct. Unlike in the United Kingdom and in Germany, the possibility of overlapping competencies has not been subject to public discussion.

6.6.2 Media policy

Media policy in France is under the strong influence of a highly protectionist industrial and cultural agenda. The main policy objectives are to strengthen the domestic program industry and to protect the media sector from foreign, especially Anglo-Saxon, influences.

To achieve this objective, terrestrial and cable broadcasters are subject to various limitations. Quotas for French and European films are imposed (60% of programming has to be of European origin, 40% of that of French origin), as well as production quotas for French films. Furthermore, the broadcasting of movies on certain days of the week (e.g., Fridays) is strictly limited to protect the cinema industry. There is also a limit on the total number of movies broadcast per annum.

With the increased penetration of direct-to-home satellite broadcasting, these program limitations lose their influence. The satellite broadcasters use non-French satellite systems and so do not fall under French law—only under the much more liberal European regulation laid down in the Television Without Frontiers Directive.

Apart from the basic aim to prevent the "Americanization" of French media, policies are highly influenced by political interests. This is shown in several changes of direction in media policy, laid down in the media laws of 1982, 1986, 1989, and 1994—always following soon after changes in government. The CSA is the only media authority to have survived several changes of government since 1989, although its structure and responsibilities are continually under review.

In general, French media policy is characterized by a fundamental contradiction. On one hand, the aim is to support the development of powerful domestic private media enterprises that are able to compete in the international environment. Concurrently there are objectives for the independence and protection of the national culture. The latter objectives require regulatory measures quite opposite to those required for the former. For example, Canal Plus has benefited from government support and light regulation or regulatory exemptions because it is a strong domestic player. At the same time, however, Canal Plus supplies the market with much U.S. content and thus works against French cultural policies that aim to protect the national culture.

6.6.3 Current media regulation

Media regulation in France is based on the 1994 media law (Carignon Law) with amendments, the latest of which came into force in July 2000. In 1994, the restriction on media ownership of private channels by single European operators had been relaxed from 25% to 49%. Non-European firms cannot hold more than 20% of a private channel in France. The liberalization was intended to allow French companies to compete on a more level playing field with competing European and U.S. groups, thereby helping to maintain the national identity of the French media. It is worth noting though that the French limits are nonetheless comparatively restrictive in comparison with German or U.K. regulations.

Furthermore, the public service broadcasters were weakened through a strict cost savings plan, laid down in December 1996, that foresaw savings of F205 million for France 2 and F20 million for France 3. This meant that France 2 had to gain more revenue from advertising and sponsorship and thus had to reorientate its programming material to mass-market tastes. France 3 was forced to freeze its rationalization strategy, which forms the main competitive advantage for this public service in the centralized TV market structure. At the same time as the public services were weakened, the private broadcasters, especially TF1, were strengthened. The CSA extended TF1's advertising slots during movies from four to six minutes, thereby allowing additional revenues of F200 to 400 million per annum [4]. Furthermore, TF1's digital venture TPS gained the exclusive distribution right for the digital broadcasting of France 2 and France 3.

Recently, the Socialist government under Jospin has aimed to reverse the deregulatory measures. The objective was to limit private media ownership and to strengthen the position of the public broadcasters. A first measure in 1998 was a 5% increase in the license fee, from F700 to F735 per annum to be paid by TV households for the reception of public service broadcasts.

A significant change in the new broadcasting bill is the obligation on pay-TV program bouquets to reserve space for third-party channels. This includes a provision not to allow the signing of exclusive contracts with public channels. As discussed in Section 6.3.2.4, this is aimed particularly at TPS, which has been exclusively broadcasting the public channels France 2 and France 3 on its digital satellite bouquet. With this privilege ceased since October 2000, CanalSatellite Numérique also transmits the public channels.

6.6.4   The impact of digital technology on regulation

The development of digital broadcasting services, characterized by a much higher number of channels and a larger role for satellite transmission, has had an impact on the regulation of broadcasting services and requires changes in the regulatory regime. The main issues to be considered as a result of the introduction of digital broadcasting services are as follows.

> • Obligations for content have already been reduced due to the fact that it appears impossible to control the content of the evolving multiplicity of channels.

> • Digital broadcasting develops mainly through satellite, which is cross-border in nature and is only regulated by minimum standards at the European level. For this reason, satellite broadcasters in France are subject to lighter (European) regulation than (French) cable or terrestrial operators. Moreover, under the European principle of "free flow of information," a national government has to permit the transmission of any channel that has been licensed in another European country. This means that French thematic channels, being subject to restrictive French regulation, are at a disadvantage vis-à-vis foreign channels that can easily gain access to the French market.

> The limits on the broadcasting of movies at certain times of the day (e.g., Friday evenings) to prevent competition with cinemas have been softened with the increasing number of channels available through digital transmission and, in particular, pay-per-view.

> Finally, quota limits have been reduced by the European Television Without Frontiers Directive. Formerly, French legislation obliged broadcasters to show 60% European content, with 40% of that French. The European directive, which overrides national legislation, demands only 50% European content, so that the requirement of French content is automatically reduced as well for non-French broadcasters. France can continue to regulate its own broadcasters more strictly.

In summary, it appears that the introduction of digital television services has occurred against a "lighter touch" regulatory background. This was certainly a factor in support of the fast penetration of digital TV in France.

6.6.5 Regulation of conditional access

In France, the 1995 EU directive [5] has been implemented as a national law but it does not provide detailed interpretation and thus remains vague. For digital broadcasters this means that, on the issue of CA, only competition law applies.

Although the CSA is in favor of a common, open CA system, it is unable to impose it on the basis of existing law [6]. It is expected that, due to rapidly falling decoder prices, a market for sales (rather than rental) will develop. The CSA assumes that companies such as Canal Plus, after initial penetration strategies, are no longer interested in providing decoders for consumers, since such a strategy carries losses. Once manufacturers sell directly to viewers, it is expected that an open (nonproprietary) CA system will evolve, since manufacturers have a strong interest in the realization of economies of scale from decoder production.

In general, it can be noted that the French regulatory authorities have taken few measures to ensure "fair, reasonable and nondiscriminatory" access for digital broadcasters, as demanded by the EU directive. In light of the two successfully competing operators (and ABSat as program supplier) established in the market, regulators do not perceive the need for intervention at the present time. The market developments so far are proving them right.

6.7 Summary

France, with its two competing platforms and a total of around three million subscribers to digital pay-TV services (13% of all TV households) within four years, was earliest and highly successful in the introduction of digital pay-TV services in Europe. Figure 6.1 gives an overview of subscriber developments in France. It is particularly notable that TPS could achieve high growth rates without having any existing analog subscribers that it could convert to digital. The cable operators have been slow to win digital customers. However, they now show stronger growth than TPS and CSN, whose growth rates are slowing down.

In particular, satellite has benefited from the move to digital (see Table 6.7). Satellite penetration has increased dramatically from 5.7% in 1997 to 18.8% in 2000. Satellite reception accounts for 87% of all digital households in France.

6.7.1 Why digital TV proved successful in France

Several factors can be identified that have significantly contributed to the success of digital TV in France. First, the existing structures in the TV

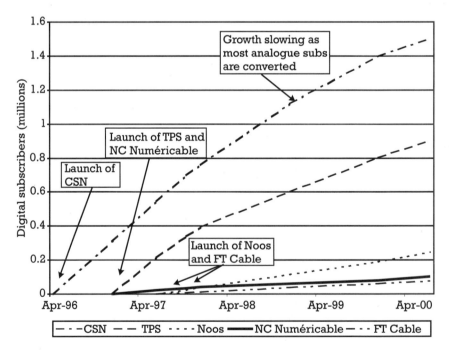

Figure 6.1 Digital subscriber developments in France, 1996–2000.

Table 6.7
French TV Households by Reception Medium, Summer 2000

	Households Analog and Digital by Medium (millions)	Percentage of Total TV Households	Digital Households by Reception Medium (millions)	Percentage of Digital Households
Satellite	4.05	18.8	2.46	87.2
Cable	2.31	10.8	0.36	12.8
Terrestrial	15.14	70.4	—	—
Total	21.49	100	2.82	100

Source: SES/Astra.

industry were supportive, in particular, the existence of a successful analog pay-TV service, the relatively small number of terrestrial channels, and the very limited penetration of satellite and cable TV. These structures meant that consumers would perceive a significant value added from new digital services providing much more choice and a better reception.

Second, the penetration strategies employed by the main players were carefully chosen. All operators launched their digital packages with a relatively large number of thematic channels and optional premium packages. This is important since potential viewers have to perceive a real advantage over existing program offerings.

Third, with the initial exception of ABSat, the necessary digital reception equipment was heavily subsidized and did not require large initial outlays from the consumer. Decoders are rented at the moderate price of F45 per month and, via promotional offers, satellite dishes were largely provided at no cost. This considerably lowers the investment threshold for consumers and reduces fears of being locked in to an incompatible system.

Fourth, the pricing of digital bouquets proved attractive. Digital bouquets are offered at a discount to analog pay TV. TPS' price differentiation strategy meant significant discounts in comparison to CSN. This also caused Canal Plus to reduce the price of some of its bouquets. Hence, the competition between two incompatible offers, leading to lower prices, had a positive impact on the take-up of digital TV.

Competition had a further effect. It has led both platforms to spend heavily on promotion and marketing so as to quickly win subscribers.

This helped to heighten consumer awareness of pay-TV offering. Moreover, new channels and services are continually being launched, and the package offers are becoming more and more flexible. This, too, is beneficial to consumers.

Finally, the experience from the French market suggests that one market can sustain two incompatible networks. The above arguments even suggest that the existence of competing platforms has enhanced the penetration of such services. The light regulation of digital services can be seen as a favorable factor for the fast penetration of digital TV. Regulation has been lighter for digital than for analog, and issues of bottleneck facilities have not received much attention and have not been regulated in France.

6.7.2 A look to the future

Recent developments in the cable industry suggest that cable will not evolve into further competing platforms but will serve as an additional means of distribution for the satellite broadcasters, addressing specific customer segments with differentiated channel lineups. The additional rollout of Internet, telephony, and possibly broadband interactive services is likely to further extend the penetration of digital services.

It can be noted, thus, that the French market has evolved into a duopoly for digital pay-TV services. With regard to future developments, the question that arises is whether this duopoly is sustainable in the long run.

At present, TPS appears firmly established in the French market. Though it has fewer subscribers, the service offers sufficient differentiation to attract viewers and is continually growing, now faster than CSN. Whereas CSN offers up-to-date high-quality content at relatively high prices, TPS focuses on less expensive, often older programs.

However, will TPS be able to compete with Canal Plus in the long run? Expenditure for content is increasing rapidly while subscription revenues are static or falling as a result of competition. Furthermore, the increasing number of thematic channels means that it becomes more and more difficult for those channels to attract an audience and, thus, be of interest to advertisers and sponsors.

Its international activities and the diversified and vertically integrated structure could form an advantage for Canal Plus in the future. It can be expected that CSN will continue to take the greatest share of digital subscribers. This is attributable to three factors. First, Canal Plus already has a strong brand name in pay TV and an existing subscriber base from which to build its offering aggressively. Second, CSN has close, well-developed links with many of the studios (French and U.S.) for pay TV and pay-per-view rights as well as long-term contracts to cover sports events. Its presence in several other European countries also means that it is a wholesale buyer of rights and can negotiate better deals. Finally, as the service with the largest subscriber base, CSN can offer studios, thematic channels, and event organizers the highest minimum guarantees and thus provides the most attractive platform for these rights owners to generate revenue. This should allow CSN to offer an overall superior package to consumers.

For TPS, the challenge will be to find and develop market segments that Canal Plus does not serve, such as interactive services, national and local programming, or special interest channels, and to serve such segments at lower prices. The connection to two large cable operators could help to achieve this.

It can be concluded that, in France, Canal Plus was successful in making use of its advantages as an incumbent and as the first mover into digital TV. It is likely that this company will stay the dominant player in the pay-TV market in the long run. TPS followed a very successful second mover strategy and, by creating competition, has supported the rapid penetration of digital TV.

The present situation is favorable to French consumers. They can benefit from network effects and, at the same time, competition ensures that they have not been locked in, prices are kept low, and offers are increasingly improved. For regulators, this implies that they should ensure that TPS stays in the market. The new legislation, removing the exclusivity of public channels on digital platforms, effectively runs against this target.

Further assessment of the developments in French digital TV, notably in comparison with other countries and with reference to the theoretical discussion in Chapter 2, can be found in Chapter 9.

Endnotes

[1]   Convergent Decisions Group, *Digital Terrestrial Television in Europe*, London, Spring 1997, p. 47.

[2]   Salomon Smith Barney, *Canal Plus Company Report* 2/98, London, 1998, p. 18.

Contents

Germany: A large market with little demand for digital pay TV

7.1   Introduction

Germany is one of the largest and most lucrative European audiovisual markets. It is also regarded as a difficult market in which to develop digital pay-TV services. Between 1996 and 1999, a number of rapidly changing alliances and strategies could be observed around Germany's largest media companies, Bertelsmann and the Kirch Group and the dominant cable operator Deutsche Telekom. So far, digital pay-TV services have failed to attract a significant number of subscribers and have resulted in heavy losses for the involved broadcasters.

Section 7.2 analyzes the specific structures of the German audiovisual market to explain why digital pay TV holds little attraction for viewers. The strategies of Kirch and Bertelsmann as well as those of Deutsche Telekom and the public broadcasters ARD and ZDF are analyzed and assessed in detail

117

in Section 7.3. Section 7.4 provides an overview of existing media regulation and examines the approaches of regulatory bodies in response to the emergence of new broadcasting services. Finally, Section 7.5 offers concluding remarks.

7.2 Market structures in analog TV

Existing market structures have to be analyzed to determine their attractiveness for consumers to switching from the current technology system (analog TV) to a new one (digital TV). This section analyzes these characteristics at the time of the launch of digital services to assess the success of such services.

7.2.1 General remarks

Germany is regarded the most lucrative European market for media due to its size (33 million TV households) and per capita income. In terms of inhabitants multiplied by per capita income, the German-speaking market is the second largest after the English-speaking market. Furthermore, Germany has a very high penetration of cable and satellite. Over 80% of TV households receive TV broadcasts via cable or satellite (see Figure 7.1). Cable is the most popular distribution medium serving more than half of all TV households. Satellite TV has shown the strongest growth, increasing from zero to 32% penetration within 10 years. Terrestrial reception now accounts for only 12% of households.

The commercial free-to-air broadcasting market in Germany is almost equally divided between two companies, the Kirch Group and Bertelsmann. Kirch controls the channels Sat1, Pro7, Kabel 1, and Deutsches Sportfernsehen (DSF). Bertelsmann has stakes in RTL, RTL2, SuperRTL, and Vox. Several other channels exist that are provided by small independent program suppliers.

Most commercial channels are incurring losses. The only profitable ones are RTL, Sat1, and Pro7. These channels have a sufficiently large viewer share to gain substantial advertising revenues (see Table 7.1). The public broadcasters ARD and ZDF enjoy high popularity due to their reputation for quality and because of their relative lack of advertising. ARD and ZDF are allowed to show advertising only between 6 p.m. and 8 p.m. and only 20 minutes per day, with no advertising on Sundays.

Pay TV is not very popular in Germany. The sole pay-TV channel, Premiere, took five years to attract a meager 1.3 million subscribers. This

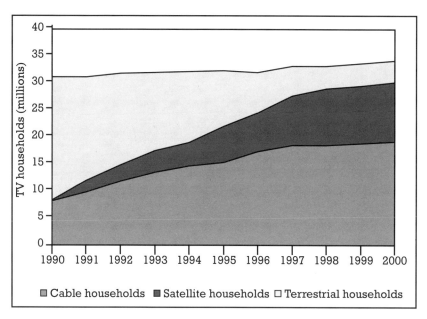

Figure 7.1 German TV households by transmission medium, 1990–2000. (*Source:* Salomon Brothers; GfK Fernsehforschung.)

represents a penetration of 4%, compared to 23% in the United Kingdom and 27% in France (1997 figures).

The fact that most German households already receive a large number of free-TV channels explains why there is little demand for even more channels. As the existing channels offer a large variety of content,

Table 7.1
Viewer Shares of the Main German Channels, 1997–1998 (%)

	1997	1998	1999
ARD	14.7	15.4	14.2
RTL	16.1	15.1	14.8
ZDF	13.4	13.6	13.2
Sat1	12.8	11.8	10.8
Pro7	9.7	8.7	8.4
Others*	33.3	35.4	38.6

*Between 15 and 20 channels.
Source: Media Perspektiven.

including movies and sports, the demand for premium pay-TV services is also rather low. In contrast, for example, to France or the United Kingdom where most households receive broadcasts terrestrially with a very limited number of free-TV channels, in Germany there is little incentive for viewers to spend extra for pay TV.

In summary, the incentives for broadcasters to introduce digital pay-TV services in Germany are not very high. Future potential may be stimulated if the number of commercial free-TV channels decreases due to competitive pressures. Already today, only three out of around 25 German channels (RTL, Sat1, and Pro7) are profitable. In addition, with TV advertising and attractive content becoming increasingly costly, some of the smaller broadcasters may be pushed out of the market or be acquired by the dominant broadcasters.

A strategy for pay-TV services would have to focus on niches such as quality premium content (the latest movies or popular sports), special interest channels, or innovative content such as interactive services.

7.2.2 A shrinking terrestrial sector

Terrestrial TV in Germany, as in France and the United Kingdom, offers only a limited number of channels (around six, depending on the location). With the high penetration of satellite and cable, the significance of terrestrial broadcasting is decreasing. Only around 12% of households watch terrestrial TV. It has to be noted, though, that terrestrial reception is still widely used for portable TV sets and in households with more than one set. The main operators of the terrestrial networks are Deutsche Telekom, the incumbent telecommunications operator, and the public broadcasters ARD and ZDF.

7.2.3 A strong cable sector

The cable industry in Germany, in contrast to the United Kingdom and France, is strongly developed and highly successful. It is dominated by Deutsche Telekom, the incumbent and formerly public service operator, which was privatized in 1996. Deutsche Telekom owns and operates the large majority (90%) of cable-TV networks up to the curb. The so-called last mile (i.e., the connection from the curb to the household), in contrast, is two-thirds controlled by private cable operators. Figure 7.2 illustrates the structure of the German cable market.

The private cable operators are dependent on Deutsche Telekom's backbone network and thus have little influence on the strategies chosen by the dominant firm. Deutsche Telekom also controls customer man-

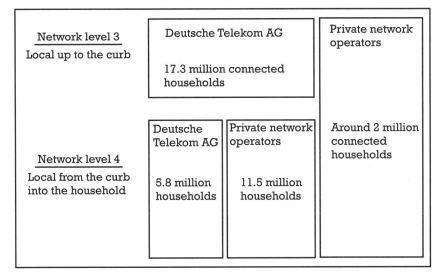

Figure 7.2   Cable-TV structure in the German market, December 1997. (*Source: Media Perspektiven*, "Basisdaten zur Mediensituation in Deutschland in 1998," Frankfurt/M., 1998, p. 9.)

agement and determines the prices for cable connection. In 1998, to watch cable TV, a household would have to pay a one-time connection charge of DM65 and a monthly fee of DM26.[1] This gives access to around 30 channels, of which 25 are national channels broadcasting in German.

7.2.4   A growing satellite sector

Satellite TV started to take off in Germany in the early 1990s and has grown steadily as satellite dishes become ever cheaper. Satellite TV is available from SES Astra or from Eutelsat's Hotbird satellites. With an Astra satellite dish, viewers receive roughly the same channels that are available on cable TV plus some additional foreign channels. Astra dishes are used in the majority of German satellite households. The Eutelsat program offer focuses on foreign-language channels and is attractive mainly to foreigners living in Germany.

In contrast to cable TV, no monthly fee is charged to receive the free-TV programs via satellite. However, the consumer has to purchase and install a satellite dish.

---

1. DM1 is equivalent to approximately $0.46.

Both cable and satellite TV are still largely independent from broadcasters (i.e., any broadcaster can rent satellite or broadband cable capacity to transmit channels). This is an important point and a factor that distinguishes the German market from the French: There is little vertical integration of commercial broadcasters and distribution networks in Germany.

7.3 The struggle of Kirch and Bertelsmann to dominate digital pay TV

Between 1996 and 2000, the two largest German broadcast companies, the privately-owned Kirch Group and Bertelsmann, a public company, fought for dominance in digital pay TV. This section looks at their respective strategies during that time.

7.3.1 Kirch's Digitales Fernsehen 1 (DF1): Pioneer in the marketplace

Kirch was first to launch a digital pay-TV service, DF1, in July 1996. However, the venture did not go quite as planned. This section explains what happened and why.

7.3.1.1 Company background: The Kirch Group

Founded in 1956 under the leadership of Leo Kirch, the Kirch Group has become Germany's leading supplier of feature films and television programming. It ranks among the five largest European film and television companies in the entertainment industry. The film library comprises more than 80,000 hours of programming: 15,000 feature films and approximately 50,000 hours of TV programming. Backed by its large program library and market experience (especially in the dealing with the large U.S. film producers) the Kirch Group has been a driving force in commercial broadcasting. In Germany, it has interests in the commercial channels Sat1 (59%), Pro7 (58%), Kabel1 (100%), Deutsches Sportfernsehen DSF (100%), and N24 (100%) and in the pay-TV channel Premiere (95%). Internationally, the Kirch Group has stakes in the Italian broadcaster Mediaset (1.3%) and the Spanish broadcaster Tele Cinco (25%).

7.3.1.2 The DF1 program offering

In July 1996, Kirch launched the first German digital pay-TV channel DF1, offering a basic package of 14 channels (later increased to 19) at a charge of DM20 per month. The offer also included 30 radio programs, an electronic program guide, the pay-per-view service Cinedom with three movies on eight channels (at DM6 per movie), and a monthly magazine. For an additional DM10 per month, two sports channels were available, containing, among other things, premier league soccer. A premium movie package (five channels) costs an additional DM15.

7.3.1.3 Transmission network constraints

At launch, DF1 could only be received by satellite, which seriously limited the potential market to around 25% of TV households that had a satellite dish. Deutsche Telekom, which serves the majority of TV households (18 million) via cable, initially refused to provide its network for digital-TV transmission (except for trials). For viewers to be able to subscribe to DF1, they had to buy a satellite dish or upgrade existing satellite equipment at a cost of around DM150, and they had to buy a d-box decoder at the subsidized price of DM890. The true decoder price was then around DM1,400 to DM1,600. The conditional access technology used in the d-box was developed by Nethold and is owned by Beta Research, a subsidiary of the Kirch Group. Kirch ordered 1 million decoders, produced exclusively by Nokia.

The high initial costs to consumers and the fear of lock-in to Kirch's proprietary decoder technology explain the observable reluctance of consumers to acquire the decoder and subscribe to DF1. It was uncertain which decoder would finally become the de facto standard in the German market. Moreover, as mentioned earlier, with a large number of free-TV channels consumers do not see an added value in paying for even more channels.

7.3.1.4 Aggressive content acquisition

For the digital services to be attractive to prospective viewers, Kirch purchased present and future output rights from Disney, MCA/Universal, Viacom/Paramount, Sony/Columbia, and Warner Bros. The company also won the $2.222 million bid for all non-U.S. rights to the 2002 and 2006 World Cup soccer tournaments. The total sum committed by Kirch for content to be shown on DF1 is estimated to be DM10 billion.

The large rights purchases from the United States serves two objectives. First, Kirch gains an important advantage over potential competitors in digital pay-TV services and can thus deter other broadcasters entering digital pay TV. Second, U.S. film companies can be prevented from entering the German pay-TV market. For the U.S. majors, Kirch has been a reliable, financially strong customer for a long time. Moreover, through contractual agreements exchanging movie rights for carriage of single channels within the DF1 program package, they can gain low-cost and low-risk access to the German TV market. This forms an ideal way of testing the German market without having to commit large initial investments.

7.3.1.5 The strategy fails

Kirch can be regarded as the classic Schumpeterian entrepreneur, who is willing to take the high risk of investing in a new market. It is obvious that Kirch has been following an aggressive first-mover strategy aimed at locking in consumers to his network. Since attractive exclusive content is the key factor for success, he acquired numerous sports and movie rights, not only to attract consumers but also to preempt the market for film rights, which has only a limited supply of top material. This is a means to keep national and international competitors out of the German market for digital pay TV, at least in the early period.

Kirch also invested huge amounts in marketing and subsidized the decoders. Most importantly, he employs a proprietary decoder technology. He thus controls several stages of the value chain for digital broadcast services: content, channel packaging, and reception equipment. Kirch has, however, no control over the distribution media, which means that access to viewers was initially very limited.

Kirch clearly aimed for a future monopoly position. Only by the prospect of high future rents can the large content investment be justified. To gain a monopoly position, he would have to lock in customers to his network. A lock-in strategy can only succeed if a large installed base can be established. By the end of 1998, DF1 had only 280,000 subscribers, half of what the company had estimated when launching the service (see Table 7.2).

Table 7.2
DF1 Subscriber Estimates for 1996–2004 at Launch

Year	End-of-Year Subscriber Numbers As Estimated by the Kirch Group	Actual Subscriber Numbers
1996	200,000	20,000
1997	900,000	50,000
2000	3 million	1.5 million*
2004	6 million	—

*Renamed Premiere World

Source: DF1 Estimates (1996).

The disappointing subscriber rates may present a substantial danger for Kirch, depending on the contractual arrangements with U.S. film producers. These contracts are based on the estimated subscriber numbers. If these are not achieved Kirch may suffer large losses both financially and in terms of credibility. In the first two years of service, DF1 is thought to have incurred losses of more than DM1 billion.

In June 1997, as a result of DF1's poor performance, Kirch modified his strategy and made the decoders available for rent. At an initial monthly rate of DM50, new subscribers receive all 30 channels. After three months, the viewer has to choose one or more of the existing program packages, which then cost between DM20 and DM40. Additionally, a monthly rental fee of DM20 was charged for the decoder. With this arrangement coming into effect, the subsidization of sold decoders terminated, increasing their retail price to DM1,200. The company's objective was then predominantly to raise subscriber numbers rapidly, rather than lock in consumers. As it became clear that DF1 was unlikely to survive by itself, Kirch aimed for a cooperation with Bertelsmann around the Premiere pay-TV channel, which was then owned jointly by both companies.

The further development of DF1 is examined in Section 7.3.3, as it now interlinks with that of the competing Premiere service.

7.3.2   The rise and fall of the Multimedia Betriebsgesellschaft (MMBG)

The second major player in the German commercial broadcasting market is Bertelsmann. This company initiated a consortium of players, named MMGB, to launch a digital pay-TV service in competition with Kirch's DF1.

7.3.2.1   Bertelsmann: A serious competitor for Kirch

Bertelsmann is one of the world's largest media groups being involved in print media, publishing, book clubs, on-line services, broadcasting, and film rights trade. Annual turnover in 1995 and 1996 amounted to DM21.5 billion with profits of DM905 million (after taxes). The television business of Bertelsmann is carried out by its subsidiary Ufa, which in January 1997 merged with Compagnie Luxembourgoise de Télédiffusion (CLT). CLT-Ufa now is the largest private television company in Europe having stakes in 19 television and 23 radio stations. Among them are the German channels RTL, RTL2, Super RTL, Vox, and Premiere.

Kirch and Bertelsmann have very different corporate cultures. Whereas Kirch, being an independent businessman who is prepared to take high risks, aggressively pushes into the market and is willing to enter into fierce competition, Bertelsmann, being a publicly listed company, has to consider shareholder interests and has therefore adopted a considerably more cautious approach.

7.3.2.2 The foundation of the MMBG

Early in 1996, Bertelsmann, together with Canal Plus and CLT, initiated the launch of a consortium for the distribution of digital pay-TV decoders. The aim was to ensure free access for all broadcasters to a CA system, known as mediabox. Free access was to be achieved through any interested broadcaster able to join the consortium. However, it has to be noted that the mediabox is also a proprietary system, not equipped with an open-interface socket, and is thus incompatible with other CA systems.

The consortium was officially launched in May 1996 under the name MMBG. Its members were Bertelsmann, Canal Plus, CLT, Deutsche Telekom, private broadcaster RTL (owned by CLT), public broadcasters ZDF and ARD, and Debis (owned by Daimler-Benz). Deutsche Telekom was the largest shareholder with a 27% stake. An 18.5% stake remained open to interested firms. Inviting new stakeholders is a means of convincing regulatory bodies that an open platform is being provided, permitting access to all broadcasters. This approach was chosen because an alliance in 1994 for the distribution of digital pay-TV services, consisting of Kirch, Bertelsmann, and Deutsche Telekom, the so-called Media Service GmbH, had been blocked by the European Commission's Competition Directorate. It was regarded as anticompetitive due to the possibility of foreclosing the market for digital pay-TV services [1].

Initially, the Kirch Group was to join the MMBG venture but after lengthy negotiations dropped out and announced that it would market and distribute its own decoder, the d-box, which is incompatible with the mediabox. Rupert Murdoch was also planning to join the alliance but also decided to drop out in June 1996 because of "unsolvable differences among the partners."

7.3.2.3 The program offering Club RTL never launched

MMBG planned to launch in October 1996 a program bouquet called Club RTL with 11 channels plus pay-per-view channels offering films in NVoD. Contractual arrangements were made, with decoders being ordered from Philips, Thomson, and Sagem, necessitating high initial investments. Contents were to be provided by Bertelsmann's subsidiary Ufa and RTL. Since there was hardly any possibility of competing with DF1 on the content side, Club RTL aimed to differentiate itself in terms of service provision: It planned to rent decoders; monthly fees were to be lower than DF1's (at a maximum of DM50 for decoder and contents); and programs were to be made available individually rather than in an all-in-one package. Furthermore, MMBG claimed to have advantages in marketing and service due to experience gained with Premiere. A considerable distribution advantage over Kirch was Deutsche Telekom's extensive cable network allowing access to 16 million households.

Club RTL, however, never took off because by July 1996, disagreements among the MMBG shareholders had arisen. With the introduction of DF1, Kirch and Bertelsmann announced that they would make their decoders compatible through the Simulcrypt method (see Chapter 5.3.3.2) and, in the future, develop a common box able to decode both types of CA systems.

This agreement was in contradiction to the MMBG arrangements since it would render MMBG obsolete , which had been set up specifically to distribute the mediabox. As a result, Bertelsmann faced strong criticism and opposition from its partners. One month later, the takeover of the South-African Nethold Group by Canal Plus, announced on 6 September 1996, changed the landscape. Canal Plus now controls both CA systems, the d-box used by Kirch, and the mediabox it created with Bertelsmann. The acquisition means that MMBG had become irrelevant as a group promoting a second standard in the German market. Canal Plus was now seen to have the power to determine which decoder would be used in Germany.

As a result of this development, Deutsche Telekom was the first to announce its intention to withdraw from MMBG on the grounds that it could become dependent on a single program provider, namely Canal Plus. It aimed to develop and market its own decoder technology. Since Deutsche Telekom has access to around half of Germany's TV households, it has a gatekeeper function concerning viewers' access. However, it has to reach an agreement with private cable providers, which own more than half of the home connections (the "last mile to the customer") and are very interested in transmitting digital-TV services.

The MMBG venture, which subsequently disappeared quietly, amounted to a costly failure for the firms involved. Bertelsmann, for example, is reported to have incurred losses of around DM230 million.

7.3.3 The pay-TV service Premiere: Competitors united

An existing (analog) pay-TV operation generally has an advantage in moving to digital pay-TV services as it has already a paying subscriber base that it can then convert to digital. Moreover, such a company has the necessary expertise in subscriber management. In Germany, Premiere, the only analog pay-TV channel with a subscriber base of 1.5 million, had these advantages.

The fundamental reason why it did not play a substantial role from the outset lies in its shareholder structure: Until May 1999, Bertelsmann and Canal Plus each held 37.5%; Kirch held the remaining 25%. Hence, the competitors in analog commercial free TV and in digital pay TV were partners in Premiere. No agreement could be reached for a long time regarding Premiere going digital. Indeed, in October 1996, Premiere instituted legal proceedings against Kirch's DF1 claiming that premium movies must not be placed in competition to Premiere but only be shown or announced after a period of 12 weeks. Ironically, it was Gottfried Zmeck, managing director of DF1 and at the same time, member of the board of Premiere, who drew attention to this issue.

Premiere was engaged in a large digital-TV trial in Hamburg, Berlin, Bavaria, and North-Rhine Westphalia, Germany, during 1997 and 1998. The digital service comprises the existing Premiere channel on three channels at staggered starting times as well as four pay-per-view channels offering mainly sports, music, and movies at DM6 per event. In these projects, the mediabox decoder technology was used. A total of 30,000 mediabox decoders have been provided: 10,000 for cable and 20,000 for satellite reception. The decoders are available to Premiere subscribers only; they must simply exchange their analog decoders for digital ones and subsequently pay a monthly rental fee of DM9.90 (increased to DM19.90 in August 1997). Monthly costs for the program package remain the same as for the analog service (i.e., DM39.90). By August 1998, Premiere had 1.65 million subscribers, of which 170,000 received the digital package.

Section 7.3.4 discusses the further development of Premiere, which is now related to that of its competitor DF1.

7.3.4 Premiere, DF1, and Deutsche Telekom: An anticompetitive alliance

During the spring of 1997, continuous rumors and speculations about an agreement between Kirch and Bertelsmann over cooperation in digital TV attracted much attention in the press. In June 1997, Bertelsmann, Kirch, and Deutsche Telekom announced their cooperation in digital TV, with Premiere being used as the main platform into which DF1 was to be merged as of January 1998. The arrangement would be a 50:50 partnership between Kirch and CLT-Ufa. The companies would buy out the Canal Plus share so that each would hold 50%. Kirch, CLT-Ufa, and Deutsche Telekom (each 33%) would jointly own Kirch's decoder company, Beta Research. The d-box was to become standard for both Premiere and DF1.

Kirch's film rights, worth around DM10 billion, would be transferred to Premiere. Premiere would also be responsible for the procurement of future content. Kirch agreed to this because, due to the huge losses of DF1, estimated at DM1 billion per year and the financial pressures of the Kirch Group, he now depended on a cooperative agreement to ensure DF1's success.

The cooperative agreement further provided that Deutsche Telekom would become the administrator of the d-box technology for cable transmission, offering access to all broadcasters. In keeping the cable network open, Deutsche Telekom claimed this would fulfill the demands of the German Federal Cartel Office (Kartellamt).

The cooperation, which appeared to be a revival of the Media Service GmbH (MSG), was subject to approval by national and European regulatory bodies. The companies claimed that the new cooperation differed from the MSG in that now the marketing of channel packages (by broadcasters) is separate from the distribution technology (controlled by Deutsche Telekom). This would allow competition of broadcasters on the basis of a single technology and, thus, would not constitute anticompetitive behavior. However, even if regulators accepted this argument for cable markets, the danger of market foreclosure would still exist for satellite markets where Kirch and CLT-Ufa control the technology.

The merger plans were submitted to the European competition authorities in late 1997. Between March and May 1998, negotiations between the three cooperating firms and the competition authorities took place. The firms agreed to make concessions so as to avoid antitrust problems. For example, regional cable operators, which strongly opposed the new alliance, were to be offered a share in the distribution of digital programs. Furthermore, Premiere offered to provide its program bouquets to competing companies for distribution. Kirch promised to open up his decoder technology to other producers and offered to sell parts of his content rights. Further, he agreed that Premiere should pass on information on its customers to competing companies—a Commission demand that Kirch and Bertelsmann had long rejected.

Bertelsmann did not back these concessions, claiming that the deal would become unprofitable for them. Hence, despite many concessions made and the lobbying of several German politicians (including then-Chancellor Helmut Kohl) to approve the alliance for the sake of a vibrant German TV sector, on May 27, 1998, the European Competition Directorate unanimously rejected the proposed merger. It was reasoned that the alliance would lead to the creation or the reinforcement of market-dominant positions that would significantly inhibit effective competition in the German market, making third TV operators dependent on the dominant firm [2].

Kirch's preparedness to make significant concessions shows his desperation to find allies to finance his loss-making DF1 venture. Immediately after the EU decision, DF1 announced that "the company still saw no rational economic basis for pursuing DF1 on its own"[3]. With this statement, the firm signaled clearly that a partner was required to carry the high costs. For Bertelsmann, the stakes were not so high. Although the company incurred considerable losses with the failure of Club RTL, digital TV was just one business plan among many profitable businesses. The decisive criterion for the company is the long-term profitability of digital TV, and with the concessions demanded by the European Commission this was not certain.

The negotiations between Kirch and Bertelsmann continued. The two players were characterized by a certain interdependence: Kirch owns extensive movie and sports rights as well as one million decoders, Premiere, in contrast, had the potential viewers. Still on the agenda in mid-1998 was the plan to buy out Canal Plus in Premiere and to form a 50:50 partnership. This venture was submitted to the Federal Cartel Office (Bundeskartellamt). The authority banned the plan in October 1998, declaring that this would lead to an oligopoly in the TV market that would be able to act as a cartel in the purchase of rights for free TV and pay TV. As such, it would be anticompetitive.

7.3.5 Kirch Group: Change of strategy

Following the disappointing takeup of DF1 and the failed alliance with Bertelsmann and Deutsche Telekom, Kirch significantly revised his strategic approach.

7.3.5.1 Restructuring and new partners

As a result of the failure of DF1 in creating a significant installed base of viewers, Kirch altered his strategic approach. As a first step, from

January 1999, the Kirch Group was restructured into three holding companies:

» TaurusFilm combines commercial free TV, license trading, program production, and film processing. During 1999, TaurusFilm was merged with the newly established Kirch Media, opening up this group to foreign investors.

» The PayCo Holding, to be merged with and renamed Kirch PayTV, includes Kirch's pay-TV activities (i.e., DF1 and Premiere).

» TaurusBeteiligung will deal with Kirch's participations in newspaper publishing, film distribution, music, and video rights, as well as software and decoder development activities.

The Kirch foundation retains a majority share in all three companies through the holding company Struktura. By separating the core activities from the largely unprofitable pay-TV ventures, Kirch can attract outside investors to his profitable business and use his own financial means for the problematic pay-TV ventures.

7.3.5.2 Opening up the decoder technology

In addition to the restructuring, Kirch ended Nokia's exclusive license for decoder production. Additional licenses were given to Philips and Sagem. Meanwhile, Betaresearch announced that it would license its technology to other producers as well. These makers would market the boxes themselves rather than the Kirch Group purchasing and distributing them. The new generation of boxes was also to be updated with regard to technology. Betaresearch decided to base new boxes on the Java programming language, the technology selected by the DVB Group for next-generation set-top boxes and idTVs and selected Sun Microsystems' ChorusOS real-time embedded operating system. These developments have significantly opened up the market for decoder production.

To ensure that the d-box would remain the only decoder standard on the German market—and in response to accusations of monopoly—Kirch made available for licensing the API Betanova, which is used in the d-box. The API license, which costs DM29,000, enables competing operators to develop their own applications for use on the d-box. Among those are EPGs, games, software for Internet access, interactive advertising, as well as market research applications. By acquiring the API license, ARD and ZDF, for example, could make their EPG accessible to d-box viewers. It would, however, first have to be converted from the presently used OpenTV standard. Allowing competitors to develop their own applications and thus differentiate from Kirch made it more likely that they would adopt the d-box rather than aiming to promote a completely different decoder.

7.3.5.3 Kirch takes over Premiere

In April 1999, Kirch announced a free-TV alliance, called Eureka, with Fininvest and Saudi Prince Al-Waleed, bringing together broadcasting and production interests in Germany, Spain, and Italy. Both partners agreed to inject DM375 million for a 3.19% stake in Kirch Media. This puts a value of DM11.7 billion on Kirch Media, which is to be floated within the next few years. Rupert Murdoch was also in talks to join the group but abandoned the plan as it became clear that Kirch's pay-TV activities, which are of greatest interest to Murdoch, would not fall under Kirch Media Holding.

Still, the three stakeholders were courting Murdoch's News Corporation to join the alliance, with the particular goal of preventing Murdoch from entering the German broadcasting market as a competitor to Kirch. This fear had intensified when Murdoch, who holds sizeable stakes in two small channels, Vox (49.9%) and TM3 (66% in 1997), purchased the rights for the European Champions League until 2003. He announced that the matches would be shown on the (former) woman's channel TM3, which has a coverage of only 80% of households and a 0.6% viewer share.

It was only in April 2000 that Murdoch, through BSkyB, took a 24% stake, worth DM1 billion, in KirchPayTV. However, there are conditions attached to this partnership. Should Premiere not prove successful (i.e., have 4 million subscribers by the end of 2002) then Murdoch would be able to exit the partnership. However, if Kirch is then unable to pay out Murdoch's stake, Murdoch has the option to acquire 51% of KirchPayTV. This would give him control over the German pay-TV venture.

As soon as Kirch had secured the deal with his new investors, he announced an agreement with Bertelsmann to raise his stake in Premiere to 95%. Bertelsmann (through CLT-Ufa) keeps 5% in order to be able to observe developments in pay TV, but stresses that it wants to refocus on its free-TV business, potentially offering such channels via digital transmission. The Bundeskartellamt permits the new solution because it effectively terminates the interlinking between Kirch and Bertelsmann in the pay-TV sector. This is seen as preventing concentration and lowering market entry barriers for newcomers. The antitrust authority accepts Kirch's monopoly position in pay TV as "not unusual in a new market with a new technology."

7.3.5.4 Relaunch of Premiere and DF1 as Premiere World

The Kirch Group now plans to invest a further DM1.7 billion in digital TV over three years, directed mainly at the acquisition of further movie and sports rights, but it also allows for an annual marketing budget of DM100 million. DF1 is merged into Premiere and the new Munich-based venture, renamed Premiere World, was relaunched with a large marketing campaign in October 1999.

Premiere World now offers four program packages, described as follows:

▸ Movie World, an offering for movie fans, costs DM39.90 per month and contains seven movie channels and the Premiere channel.

▸ Family World offers a selection for the whole family but with no movies. At DM19.90 per month, the package offers four children's channels (including Fox Kids and the Disney Channel), two channels for soap operas and series, and two documentary channels.

▸ Gala World targets the older generations, offering six thematic channels including movies, classical music, documentary, and series for DM19.90 per month.

▸ Sports World offers premium sports channels. The "light" version, at DM29.90, offers three sports channels, but without soccer. Sports World Kick is the full offering. At DM39.90 per month the viewer gets five sports channels, including several live Premier League, Champions League, and international soccer games. This offering also allows access to Superdom, the pay-per-view service for all Premier League soccer games. At DM12 per day, DM25 per week, or DM349 for a whole season, viewers can choose the live games they want to watch.

Additional premium channels can be purchased either separately or in combination with a package, then at a discount. Those available are the following.

▸ The opera channel Classica at DM20 (separately) or DM10 (in combination with a package);

▸ The documentary channel Seasons at DM15 (separately) or DM10 (in combination with a package);

▸ The adult channel Blue Channel at DM20 (separately) or DM14 (in combination with a package).

The pay-per-view services Cinedom (for movies) and Blue Movie (for adult movies) offer a discount for subscribers of the Movie World package. Films cost DM6 for subscribers and DM10 for nonsubscribers.

The sports package Sports World could initially only be purchased in combination with Movie or Family World. This was to help the Kirch Group to refinance costly sports rights. As subscriber numbers remained disappointing, however, the strategy was altered. Since early 2000, Sports World can now be subscribed to separately. The price difference between Sports World only and Sports World plus Movie World is small at DM5. It can therefore be expected that most customers will still purchase both packages.

In general, the pay-TV packages are still relatively expensive. Subscribers pay DM14.90 per month for decoder rental and a one-off connection fee of DM29.90 and face monthly program costs in the range of DM19.90 to DM54.90. Premiere World claims that 90% of its customers subscribe to all four packages at DM54.90 per month, hence paying a total of DM840 per year. Soccer fans subscribing to a Premier League season actually pay around DM1,000 per year.

The service aims to offer mainly exclusive content, so as to differentiate itself from free-TV offers. To that end, Kirch has cancelled several contracts with channels that are available elsewhere at no charge. Among those are MTV Germany, VH-1 Germany, NBC, CNBC, and BBC Prime.

7.3.5.5   Still few subscribers for Premiere World

As of September 2000, Premiere World had around 2.2 million subscribers, of which 1.6 million are digital. That represents 4.7% of total TV households in Germany.

The stake in Premiere cost the Kirch Group around DM2.74 billion. DM1.6 billion was paid to CLT-Ufa for its 45% stake. Before, Kirch had already paid DM780 million to Canal Plus for its 25%, and Kirch's share of the Premiere start-up losses amounted to around DM360 million.

Together with the start-up losses of DF1 (around DM1 billion) and the newly planned investments of DM1.7 billion, Kirch will have spent around DM5.4 billion on launching digital pay TV. This represents an immense investment in the development of a new network. Kirch's preparedness to carry these costs and his persistence in developing this market illustrates his conviction that digital TV will be a highly profitable market in the future.

An important improvement to Kirch's earlier attempts to establish digital pay TV is that he has now secured the support of Deutsche Telekom. The cable operator has adopted the d-box as the only decoder for its cable customers, and Kirch has rented 54 channels on the cable network. Furthermore, the two companies have announced a marketing agreement under which Deutsche Telekom sells d-box decoders (at DM449) and Premiere World subscriptions to its cable subscribers through its retail shops. In exchange, Deutsche Telekom will receive a set commission on subscription sales. With access to the cable customers and with the support of the large, well-branded cable operator, the conditions have improved for the pay-TV operator. With regard to the programs, pricing, and consumer choice, however, it appears that not much has changed.

7.4   Public broadcasters ARD and ZDF provide digital free TV

Public broadcasters ARD and ZDF have adopted a very different approach to that of Kirch and Bertelsmann. They plan to introduce digital television not via pay TV, but through the existing free-TV channels. With services free of charge and advertising revenues for public broadcasters being static, financing of this new service will have to be made through TV license fees. Public broadcasters in Germany are not allowed to provide pay-TV services.

Public broadcasters are highly popular in Germany, and the introduction of digital services could further enhance their market share. They have, however, so far not been able to offer services with a clear differentiation from analog services. This can be attributed to their limited possibilities to finance such ventures: With a fixed budget from license fees and without the possibility of offering pay-TV services, they had no incentive to push the transition to digital. Unlike its counterpart in the United Kingdom, the German government has done little to encourage public broadcasters to start digital transmission. Support from the government could help to quickly facilitate the transition from analog to digital. The lack of this support is another reason for the slow start of digital services in Germany.

Initially ARD and ZDF were to develop their own decoder, which would merely transform digital signals into analog, hence not including CA. This idea was subsequently abandoned. ZDF, CLT-Ufa, and Deutsche Telekom agreed to the d-box becoming the de facto standard in Germany. ARD, in contrast, had always opposed Kirch's proprietary set-top box. ARD had developed its own EPG, so as not to be disadvantaged by Kirch's EPG. However, the ARD application could not run on d-box decoders. Deployed are boxes supplied by Panasonic, Galaxis, and Kathrein, costing around DM1,000. The OpenTV API is used in these decoders. There is no in-built CA but a common interface slot so that the boxes can be upgraded for pay TV. These boxes are being used by the free-TV venture Free Universe Network (FUN), which was established by ARD together with the private cable operators.

7.4.1 ARD Digital

ARD's strategy for digital TV is not so much to provide new programs, but to create content-specific connections between the existing public channels through a navigator system (EPG), time-staggered repetitions, and the possibility for viewers to tailor their viewing through a series of "bookmarks."

ARD can build on its large number of existing channels, including its main national channel Das Erste, eight regional channels, the Bavarian educational channel BRalpha, and four channels that are run in partnership with ZDF or other broadcasters. Those are 3Sat (culture), Arte (European cultural program), KiKa (advertising-free children's channels), and Phoenix (news and documentaries). Furthermore, ARD operates 22 radio channels.

ARD Digital offers three new, digital-only channels:

- EinsMuXx: A time-staggered version of the ARD national channel Das Erste;

- EinsExtra: Supplementary information with documentaries, news, and features;

- EinsFestival: TV films from the ARD archives.

An important component of ARD Digital is the EPG, which guides viewers through the ARD family of channels. In particular, the bookmark function allows viewers to mark topics of their interest. All programs or information available on this topic will then be listed for the viewer.

Another interactive service, the ARD Online Kanal, a data and video Internet TV channel, provides information from ARD's videotext and Web sites.

7.4.2   ZDF.vision

ZDF's digital offer ZDF.vision was launched in August 1997, initially as a trial simulcasting existing free-TV channels. In contrast to ARD, ZDF has significantly fewer channels. It operates the main national ZDF channel and those run in partnership with ARD and other broadcasters, namely, 3Sat, Kika, Arte, and Phoenix.

ZDF aims to gain further channels for its digital package through cooperative efforts. Since April 2000, German broadcast law permits public broadcasters to extend their program offerings with channels from other (public or commercial) broadcasters, as long as those are licensed somewhere in Europe and fit into the existing package.

So far, ZDF has struck partnerships with three "invited channels":

▸ ORFsat: Generalist channel from the Austrian public broadcaster ORF;

▸ EuroNews: European news channel in six languages;

▸ Eurosport: European sports channel with versions in several languages.

In addition to those partnerships, ZDF has created four new channels:

▸ ZDF.info: Information on lifestyle, business, science, health, and regional news;

▸ ZDF.doku: Launched in April 2000 as an extension to ZDF.info, providing reports and documentaries;

▸ ZDF Theaterkanal: Launched in December 1999, this channel screens theatre recordings.

A simple interactive service is provided in ZDF.digitext. It offers advanced videotext functions and a search function and can display photos and graphics. ZDF.vision also comprises an EPG. ZDF uses the d-box, which cannot provide advanced interactive services.

7.5 Deutsche Telekom: Becoming an independent service provider

Having been blocked from an alliance with Kirch and Bertelsmann, the telecoms and cable operator Deutsche Telekom decided to become a digital service provider in its own right. This section illustrates that Deutsche Telekom could evolve as a major player in digital pay TV as it has (and will continue to have) significant control over Germany's cable network.

7.5.1 Deutsche Telekom's cable network restructuring

In 1998, under pressure from the EU competition authorities for its dominant position in TV distribution, Deutsche Telekom restructured its cable TV business, separating the nationwide network into nine regional operations. The cable network is expected to be highly profitable in the future. With access to 51% of households, much higher than in most other countries except the United States (see Figure 7.3), the German cable network has immense potential for new digital broadband services. After modernization investments estimated at DM3 billion to DM5 billion, turnover is expected to have quadrupled to DM10.2 billion by 2005. At that time, the cable network is expected to supply around 19 million

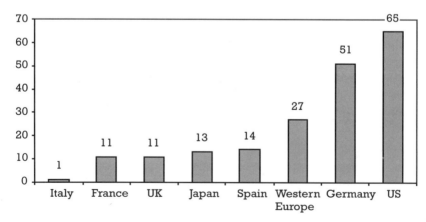

Figure 7.3 Cable-TV connections in selected countries, 1998 (%). (*Source: Die Zeit*, No. 20, December 5, 1999, p. 26.)

customers with more than 100 TV channels, high-speed Internet, and telephony.

The cable operator decided in April 1999 to invite external investors to share the high initial investments. The company's positive outlook interests, for example, Bertelsmann and Microsoft, but also tele-communications operators and some large German banks that would resell their stakes (with a profit) once the upgrading is achieved. The U.S. company Callahan and the United Kingdom–based Klesch Group, for example, have acquired majority stakes in three regional cable networks.

A contentious issue between Deutsche Telekom and potential inves-tors is the former's insistence upon retaining a 25% stake in each of the regional networks. This effectively gives Deutsche Telekom a blocking stake with which to veto a partner's efforts to offer services that threaten Deutsche Telekom's existing business, and to control the pace of consoli-dation. A further uncertainty for investors is the unclear regulatory situa-tion (see also Section 7.4). With telecommunications being regulated by the federal government and cable TV by the states (Länder), the responsi-bility for broadband cable services is not clearly allocated. The fear is that the Länder regulators could exercise an excessive influence in line with their own objectives.

The aim of Deutsche Telekom in the transition to digital TV had always been to stay independent of the large broadcasters, so as to decide freely which programs to transmit over its extensive cable network. It has to be noted, though, that the allocation of multiplexes (channel frequen-cies) on the cable networks is subject to regulation by the media authori-ties. At present, 13 cable multiplexes are allocated to digital broadcasts, with three of those reserved for the public broadcasters. With each multi-plex able to carry nine digital-TV channels, there is capacity for over 100 digital channels. The allocation of channels on these multiplexes is in the hands of Deutsche Telekom.

Deutsche Telekom has allocated five multiplexes to Premiere World; two are used by ARD and one by ZDF. A further multiplex is reserved for regional and local broadcasts. The remaining four multiplexes are used by Deutsche Telekom's subsidiary MSG MediaServices.

7.5.2 Deutsche Telekom's program offering MediaVision

Under the overall brand MediaVision, MSG MediaServices operates a digital program platform, based on the d-box decoder, with five program packages:

‣ VisionBasic (free TV) comprises the ARD Digital and ZDF.vision packages, two shopping channels, and the lifestyle channel SingleTV.

‣ VisionSelect (pay TV) offers 10 thematic channels at between DM9.90 and DM24.90, depending on the number of channels chosen.

‣ VisionGlobe and VisionSpecial target German immigrants. As Germany has a large foreign population (7 million), there is significant potential in this market segment. VisionGlobe offers Turkish-, Chinese-, Greek-, Portuguese-, and Polish-language channels. All are free of charge. Under VisionSpecial, Deutsche Telekom offers an Indian and a Russian generalist channel at DM36 and DM29.90 per month, respectively.

‣ VisionProfi, comprising a music and fashion channel and Bloomberg's business news channel, is offered to shops and restaurants at a monthly subscription of DM50 to DM100, depending on shop size.

An overview of all available channels on MediaVision is provided through the EPG, which is called the Vision Program Manager.

This new role of Deutsche Telekom as a platform operator and program packager highlights an important new development in the German broadcast sector: In the past, distribution networks were separate from content provision. Now, Deutsche Telekom and other cable operators, such as PrimaCom (see Section 7.3.8.6), are starting to determine what programs and packages are delivered to viewers. Deutsche Telekom does not reveal subscriber figures for MediaVision, and it can be expected that the customer base is still small; however, with access to the majority of German households, Deutsche Telekom could become a major player in pay-TV provision.

7.6 Plans for further digital program offerings

In Germany, there are several other broadcasters planning to offer digital packages, either as free TV or pay TV. Among those are Bertelsmann's RTL Television, music channels MTV and Onyx, independent venture @TV, Dutch cable operator UPC, and German cable operator PrimaCom.

7.6.1 RTL Television

In March 2000, RTL Television was awarded licenses for four free-to-air thematic channels covering series, news, action, and highlights plus an EPG. This program package, named RTL World, and earmarked for satellite transmission, will use decoders with OpenTV technology and will hence be incompatible to the d-box.

The Kirch Group, which has a similar program family with Sat1, ProSieben, Kabel1, DSF, and N24, has so far not announced any plans for digital free TV. Such an announcement is, however, expected soon.

CLT-Ufa, partly owned by Bertelsmann, also applied for three cable multiplexes, aiming to offer its free-TV channel family (RTL, RTLII, SuperRTL, and Vox) plus additional channels (news, weather, and shopping) as a digital free-TV package. CLT-Ufa is also very interested in developing broadband interactive TV services and Internet via TV.

7.6.2 MTV Networks

Music broadcaster MTV Networks received licenses for two digital music channels in August 2000:

‣ MTV Base: Featuring rhythm and blues, hip-hop, and dance music;

‣ MTV Extra: Featuring rock and independent music.

‣ These channels are available already via satellite in the English language. A German-language version is planned.

7.6.3 Onyx Television

The music channel Onyx received licenses for five digital music channels as well as a digital version of the existing analog Onyx channel. The broadcaster also announced an entertainment package, Onyxplus, comprising 12 digital pay-TV channels. This was planned for August 2000 but has not started yet.

7.6.4 @TV

In 1999, @TV, an independent program provider, announced the launch of 10 thematic pay-TV channels as well as 14 pay-per-view channels with movies, sports, and events. However, due to financial difficulties, this venture has still not come off the ground. The investments needed to acquire content proved too high. This case illustrates the difficulties facing new entrants wanting to offer digital services.

7.6.5   United Pan-Europe Communications

Dutch cable operator United Pan-Europe Communications (UPC) is following an aggressive pan-European expansion strategy. In Germany, it plans to launch up to 80 digital cable channels as well as telephone and Internet services. This pay-TV offering is to be positioned as an alternative to Premiere World and will not use the d-box decoder. However, UPC will have to get access to the cable network to offer these services. UPC's usual strategy is to acquire regional cable networks. As Deutsche Telekom retains a minority stake in each of its cable networks sold, it is unlikely that UPC can succeed in this way. Another way is to cooperate with the independent cable operators in Germany. This, however, will make it difficult to quickly reach a large audience. At present, UPC cooperates with PrimaCom to bring its channels to German viewers.

7.6.6   PrimaCom

PrimaCom AG is the third largest cable operator in Germany with 1.36 million connected households. PrimaCom aims to offer a digital platform with a focus on interactive services. To this end, PrimaCom is upgrading its regional cable networks for broadband capability. By the end of 2000, around 200,000 households are said to be broadband-enabled. PrimaCom plans to offer e-commerce services as well as high-speed Internet, VoD, pay-per-view, and Internet telephony.

PrimaCom is currently running trials in 700 households, offering 80 TV channels in five digital packages at DM5 to DM13 and pay-per-view at DM6 per film. The cable operator cooperates with a range of content providers to be able to offer attractive content. Among its partners are UPC, AB Groupe, and MCM International (a joint venture between the French media group Lagardère and Canal Plus).

7.7   DTT: Trials for a small market

With only around 12% of viewers receiving terrestrial broadcasts, the attractiveness of DTT in Germany is rather limited and is confined to technology trials carried out by the transmission companies Deutsche Telekom and ARD. These are to be extended gradually, but a significant commercial launch is not expected before 2004.

Due to the high cable and satellite penetration, DTT is being mainly explored for mobile reception in cars and trains. A recent development program by Nokia, Deutsche Telekom, and ZDF is testing the so-called

MediaScreen device. This combination of a TV, an Internet access system, and a mobile phone is able to receive DTT transmissions while on the move as well as offer interactive mobile functions. It can be expected, though, that DTT will not become a driving force for the introduction of digital TV in Germany. It will rather have to be pushed by the government as a means to fulfill universal service obligations. So far, however, the German government has not been active in this direction.

7.8 Media regulation in Germany

7.8.1 A complex system with state regulation of media

The German broadcasting sector has traditionally had few providers. Service provision is costly and hence, market entry barriers are high. With the ongoing process of concentration in the audiovisual sector, free market access and plurality of opinion have to be safeguarded through regulation. Public broadcasters are regulated by bodies incorporated in their organizational setup, similar to the BBC's Board of Governors. Private commercial broadcasters are regulated by external authorities.

The regulation of broadcast and new media services in Germany is subject to specific problems due to the country's federal structure. Broadcasting is classified as part of the nation's cultural life, and cultural matters are, under the German Constitution (Grundgesetz), the concern of the Länder (states), not the federal government. Consequently, each of the 16 Länder has a separate, independent media authority, Landesmedienanstalt (LMA), which is responsible for granting licenses, controlling content, and preventing concentration of media power. Only those of Berlin and Brandenburg have merged; hence, there are 15 LMAs in Germany.

The National Convention on Broadcasting (Rundfunkstaatsvertrag) is an agreement between the Länder for the purpose of assimilation of matters that are within their legislative competence. It determines that private operators need a license to arrange and transmit radio and television broadcasts. The licenses are granted and controlled by the Landesmedienanstalten. Although theoretically each one is responsible for granting a local license for a new TV channel in its region, the others will automatically approve such a grant, and effectively a national license is granted.

The control of media concentration is one of the responsibilities of the LMA. The 15 LMAs organize a working group (Arbeitsgemeinschaft der LMA) for cooperation. This working group, in turn, establishes a regular meeting of the LMA directors (Direktorenkonferenz) as well as 12 cooperative offices. This organization of media regulation is very costly. In 1999, DM223 million of the license fee revenues was allocated to the LMAs, with their 400 employees. This includes resources that the LMAs provide as subsidies for technology, programming, education, and film production for the private film and TV sector. The LMAs are widely criticized for their large budgets and above-average salaries (in comparison to similar public-service positions).

In addition to the LMAs, several other bodies also share their responsibilities: the antitrust authorities of the Länder (Landeskartellbehörden), the federal antitrust authority (Bundeskartellamt), the European Commission's Competition Directorate, as well as the 1997-launched Kommission zur Ermittlung der Konzentration (KEK)—a commission for the investigation into concentration. Furthermore, with increasing convergence between media and telecommunications, the federal telecommunications regulator (Regulierungsbehörde für Telekom und Post) is involved as well. The multitude of regulatory authorities with overlapping responsibilities and differing objectives creates uncertainty for corporations and is claimed to inhibit innovation and competitiveness on the German market. Examples are the unclear legal position of TV broadcasting via Internet or telephony services via broadband cable. Moreover, regulatory processes are complex and inefficient and often involve a duplication of work.

Since almost all of the commercial broadcasters in Germany operate nationwide, Länder regulation today is rendered obsolete. Many commentators suggest that a national regulatory body would be more efficient. However, this is politically impossible since the role of the Länder is fixed in Germany's constitution. A more realistic option would be the merger of the LMA into one single, nationwide body. The transfer of responsibilities from the LMA to the Direktorenkonferenz can be seen as a step in this direction.

7.8.2 Control of media concentration

The control of media concentration is specified in the Rundfunkstaatsvertrag, a law that limits the concentration of private national broadcasting media as a precaution against the predominance of one or a few opinions. Historically, a private broadcaster was allowed to run two nationwide television channels, but not more than one so-called full channel (Vollprogramm), which means a generalist service covering the cultural, information, and entertainment spheres comprehensively. This is because the legislators estimated the influence of those services on the public opinion to be very high. Furthermore, a private operator should not own more than 49.9% of a "full channel" and no more than 24.9% of any additional channel.

Since January 1997, new ownership rules have been based on a fundamentally different concept, namely the audience-share model [4]. A single broadcaster can now own an unlimited number of channels up to 100% so long as it does not gain an audience share of more than 30%. The full audience share is ascribed to the owners of a channel if they hold at least 25% of its stock. If a channel reaches 10% of the audience it would have to open its schedule for the programs of third parties. The new controlling authority KEK has been set up to enforce compliance with these regulations.

Critics of the new concept argue that it offers a free ticket for the expansion and further concentration of media companies. None of the existing media groups will ever reach the 30% limit since public broadcasting (ARD and ZDF) is always included in the calculation. Together, these account for approximately 40% of the national audience. Bertelsmann's (RTL, RTL2, Super RTL, Premiere) audience share comes to about 25%, the Kirch Group's (Sat1, Premiere, Pro7, Kabel1) to just under 30%. Even if a company should reach a 30% audience share, it would not have to dismantle its holding if it were able to disprove the presumption of a "predominant opinion."

Finally, following the new legislation, mergers such as Bertelsmann's subsidiary Ufa with CLT or the cooperation between Kirch and BSkyB become possible. Under the old ownership rules they would have been illegal. It can be expected that the tendency will be for concentration in broadcasting markets to accelerate as a result of this deregulation of media ownership.

7.8.3 The implications of convergence for regulation

While media are regulated at the state level, the federal government has exclusive competence in matters of telecommunications. This means that there is a distinction to be drawn between the content of broadcasting and the technical aspect of transmission. With the emergence of new services and converging structures combining telecommunications and media, this distinction has to be reconsidered. In June 1996, the first draft of a Multimedia Act (Informations und Kommunikationsdienstegesetz) proposed to introduce a regulatory framework for the free development of new services in the multimedia sector [5]. It includes provisions on free access, responsibilities for content, and effective protection from misuse of new services. The draft is currently restricted to very basic regulatory provisions. An over-detailed draft is regarded as a potential hindrance to the fast development of a viable regulatory framework and, moreover, is likely to delay the economic take-off of new multimedia services. The draft was proposed to the German parliament in December 1996 and the final act was passed in July 1997.

At the same time, in cooperation with the federal government, the Länder developed a Media Services Act (Mediendienstestaatsvertrag) [6]. In the new draft, two new categories, "teleservices" and "media services" are introduced in addition to the existing ones, telecommunications and broadcasting. The federal government, in addition to telecommunications, will be responsible for teleservices. These are defined as "all electronic information and communication services [that] allow individual use of text, images, or sound transmitted via telecommunication." Examples are e-mail, on-line services, telebanking, teleworking, or telelearning. The Länder, in addition to broadcasting, will regulate media services, defined as teleservices, that are addressed to the public at large.

Teleservices and media services are distinguished according to the group of users. If only a limited number of people, such as telebanking customers and certain professional groups, are addressed, this falls under the teleservice category. In practice, problems arise in the exact allocation of services to these categories. Several types of services, teleshopping, for example, are addressed to the public at large and, at the same time, permit individual usage of text, image, and sound. It has to be examined whether these categories actually require different legislative handling. It is more likely that this differentiation merely reflects a political compromise between the Länder and the federal government.

7.8.4 Rundfunkstaatsvertrag 2000: A proactive role for the government

The first government initiative actively supporting the development of digital broadcasting in Germany was launched in September 1998. It is laid down in the new broadcasting bill (Rundfunkstaatsvertrag) that came into force in April 2000.

Most importantly, the government determined 2010 as the year for termination of analog broadcasting. (This date is to be reviewed in 2003.) The announcement was positively received by industry as it creates certainty and an incentive for broadcasters and telecommunications operators to invest in digital TV.

To strengthen the role of public broadcasters in the transition to digital TV, the new broadcasting bill guarantees access to three digital cable multiplexes, offering capacity for more than 30 TV channels. ARD and its regional affiliates have been allocated two multiplexes; ZDF has received one. This provides an incentive for the public broadcasters to create new channels and to encourage the transition to digital. Public broadcasters are still not allowed to offer commercial services. However, they can now "invite" channels from other (public or commercial) broadcasters into their own program packages. Conditions are that these "fit into" the package and that they are clearly recognizable as channels from external providers.

With the aim to support the development of niche channels, the new broadcasting bill obliges cable operators to devote one-third of their digital capacity to the creation of media variety. This is intended to prevent cable operators from simply simulcasting the existing mainstream channels.

One important aim of the new broadcasting bill was to be more pro-active with regard to open access regulation. The new law stipulates that digital set-top boxes must be able to receive all digital channels. In June 2000, the LMA directors agreed on a statute for free access to digital services. This statute, however, does not mandate a common interface, but merely names the common interface as "one possibility to create open, nondiscriminatory access to digital services." This, in fact, means that proprietary, incompatible decoder technologies will continue to be used until the DVB MHP is introduced. All market players, including the Kirch Group, Bertelsmann, ARD, and ZDF, have agreed to employ the Multimedia Home Platform in the future.

It can be concluded that this new media law constitutes the first active attempt of the government to provide a regulatory framework for the provision of digital services. The regulations aim to ensure a role for the public broadcasters and to safeguard market access of independent third broadcasters to digital cable networks. As was argued in Section 7.3.6, the public broadcasters could play an important role in the fast transition to digital; however, they require government support. With regard to open, nondiscriminatory access to digital services, however, the broadcasting bill has not advanced.

7.9 Summary

In Germany, digital TV has not attracted many viewers. Within four years, only around 1.75 million households, 5% of total German households, adopted digital TV. Of those, 1.6 million subscribed to the digital pay-TV service Premiere World; the remaining 150,000 watch digital free TV or have subscribed to one of the smaller pay-TV offerings. Figure 7.4 illustrates the digital subscriber developments over time.

7.9.1 Why digital pay TV proved unsuccessful in Germany

The developments in Germany over the past years signify that the market for digital pay-TV services will have to be created; it will not develop endogenously. A differentiation from free-TV services (i.e., attractive content) is crucial to achieve mass penetration. Experience suggests that consumers react in a confused manner to competing technological systems when differences in value are hard to identify. This results in a reluctance to purchase any system, thus delaying mass-market penetration.

It is very likely that the competition between Kirch and Bertelsmann is at least partly responsible for the low consumer interest in digital pay-TV services. In contrast to France, where competition has led to high consumer awareness, better services, and lower prices, in Germany the competition between two incompatible systems has raised consumer fears of lock-in and concerns of abuse of a monopoly position of a dominant player. Other influential factors for the slow take-up of digital TV in Germany are the large number of free-TV channels already available, a

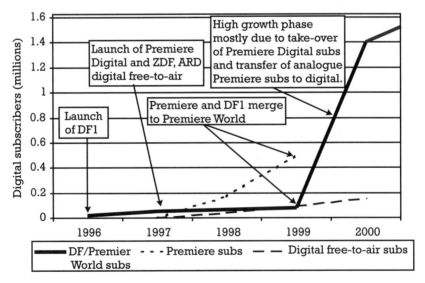

Figure 7.4 Digital subscriber developments in Germany, 1996–2000.

general apathy on the part of consumers to adopt a completely new system, and the relatively high initial outlay.

The fact that all players adopted and subsequently modified or abandoned their market entry strategies reflects the high degree of uncertainty about the best way of developing these new markets. The establishment and disintegration of cooperative agreements illustrates the difficulties in choosing an appropriate market strategy.

7.9.2 Kirch's strategic mistakes

Kirch was quickly able to identify the strategic factors in the introduction of digital television. It appears, however, that he has overestimated his first-mover advantage. DF1 did not offer sufficient added value to convince consumers to subscribe. Moreover, his strategy concentrated on the provision of attractive content. He failed to ensure access to as many customers as possible (notably via cable) and did not have an installed base of (analog) pay-TV customers, which can typically be more easily converted to digital.

Once it became apparent that Kirch's initial strategy could not succeed, he aimed to get access to the cable network and the analog pay-TV base through cooperation with Deutsche Telekom and Bertelsmann. This was, however, blocked by competition authorities. Three years after the first digital launch, Kirch relaunched, after having had to acquire Premiere. However, the available packages are relatively expensive and offer few improvements in terms of viewer choice.

Interactive services and Internet access via TV may have acted as drivers, but there are few interactive services available. This is due to the d-box, which is widely regarded as outdated, being the predominant decoder technology.

7.9.3 A look to the future

Competitive pay-TV offerings are as yet marginal, and it is unlikely that effective, large-scale competition will arise from regional cable operators, especially since Deutsche Telekom is keeping a blocking minority stake in all cable networks it is selling. As Deutsche Telekom is operating closely with Kirch in pay TV, the cable company would make sure that competitive offers cannot be launched. The smaller cable operators will only be able to compete on a regional scale.

Pay TV is still the main driver for penetration of digital TV, but it is also developing through free TV with the public and commercial broadcasters gradually moving their offers to digital and adding complementary services. This is, however, developing slowly, as these strategies do not result in significantly higher future revenues. Hence, the companies are not prepared to heavily invest in the move to digital. By October 2000, only around 150,000 households received free-TV-only digital programs.

The development of the regulatory environment for new services will also prove critical. German authorities are aware of the need to adapt regulation to newly developing multimedia services. The new Rundfunk-staatsvertrag contains several measures positive for the development of digital TV. However, there are no initiatives as yet to alter the inadequate pre-existing regulatory structures. Those have created uncertainty in the past and are likely to have delayed the transition to digital.

Further assessment of the developments in Germany, notably in comparison with other countries and with reference to the theoretical discussion in Chapter 2, can be found in Chapter 9.

Endnotes

[1]   Commission of the European Communities, "Entscheidung der Kommission im Fall MSG Media Service GmbH," *Wirtschaft und Wettbewerb*, EV, 1994, pp. 2,231–2,264.

[2]   Commission of the European Communities, "Entscheidung der Kommission vom 27.5.1998, Sache Nr. IV/M.993 Bertelsmann/ Kirch/Premiere," K(1998)1439 endg., *Media Perspektiven Dokumentation II*, 1998, p. 46.

[3]   Kirch Group, *New petition still possible*, press release, München, 27.5.1998, p. 1.

[4]   Länderregierungen, "Dritter Staatsvertrag zur Änderung Rundfunkrechtlicher Staatsverträge," *Media Perspektiven*, Dokumentation I, 1996, pp. 1–34.

[5]   Bundesministerium für Bildung, Wissenschaft und Forschung, *Gesetz des Bundes zur Regelung der Rahmenbedingungen für Informations und Kommunikationsdienste*, Referentenentwurf, Bonn, June 28, 1996.

[6]   Länderregierungen: *Staatsvertrag über Mediendienste*, August 1996.

CHAPTER 8

Contents

The United Kingdom: Late introduction and fast penetration of digital TV

8.1 Introduction

The United Kingdom has moved to the forefront of technological developments for digital-television services in Europe, and a powerful, international pay-TV operator, BSkyB, is established in the market despite the fact that the United Kingdom is one of the last countries in Europe to introduce such services. Only from October 1998 were digital-TV services rolled out across the country.

Moreover, the United Kingdom was the first country in Europe to launch DTT with public and commercial operators offering free-to-air programming as well as pay TV. Compared to other European countries, where different digital services and distribution media were launched over time, the rollout in the United Kingdom has been more or less a "big bang." All providers and all distribution media have been launched

151

within less than one year. Digital satellite launched in October 1998, followed by DTT in November 1998, with digital cable services commencing in July 1999.

This chapter analyzes the developments prior to the launch of digital services and their success since then. The examination of existing market structures and the strategies of the digital-TV providers allow some conclusions to be drawn about the longer-term prospects of the different digital services in the United Kingdom.

Section 8.2 gives an overview of the existing market structures in (analog) terrestrial, cable, and satellite TV. Section 8.3 examines the development and planned strategies for DTT. Section 8.4 assesses the strategies of the main commercial competitors, and Section 8.5 looks at the opportunities and plans of cable operators in the digital era. Section 8.6 focuses on regulatory approaches in the United Kingdom, and finally, Section 8.7 offers some conclusions.

8.2   The structure of the U.K. television market prior to digital launch

With a situation similar to that of France, the United Kingdom also depicted marked structures conducive to a succesful introduction of digital pay-TV. Most notable are a small terrestrial offer for the majority of households, a struggling cable industry, and an established pay-TV operator. This section looks in detail at the U.K. broadcast market for analog TV.

8.2.1   A small terrestrial offering for the large part of the market

In 1998, prior to the launch of digital TV, more than two-thirds of the 23 million U.K. television households only received the terrestrial program offering of four to five channels. The British Broadcasting Corporation (BBC), a public broadcaster established in 1927 and financed through license fees, provides two of those channels, BBC1 and BBC2. The remaining three channels, ITV, Channel 4, and Channel 5, are commercially funded.

Commercial TV, financed through advertising, was first introduced in the 1950s on the basis of a federal system of regional areas. The national channel ITV is provided by 16 regional ITV companies supplying partly nationwide programs and partly regional variations. A second commercial channel, Channel 4, was established in 1982, for the purpose of further differentiating the audience and serving minority interests. In April 1997, Channel 5 was launched as another general-interest channel.

Until the 1980s, the BBC was highly vertically integrated, controlling the creation, packaging, and distribution of its programs. The Independent Broadcasting Authority (IBA), which also acted as regulator for the ITV companies, operated the infrastructure for ITV. Only Channel 4, in order to cater to minority interests, chose a different structure, using independent program providers but utilizing the ITV transmitters for distribution. In recognition that a vertical structure was not necessary and undesirable, the 1990 Broadcasting Act (for commercial broadcasters) and the renewed 1996 Charter (for the BBC) provided for the sale of the ITV and BBC transmission networks to private companies (Castle Transmission and NTL, respectively).

As illustrated in Table 8.1, the popularity of British public and commercial broadcasters is about equal, and audience shares have been relatively stable over time. Among the five main channels, BBC1 and ITV have the highest audience shares with around 30% each. BBC2 and Channel 4 (and its Welsh equivalent S4C) reach around 10% each. Channel 5 has a lower audience share of about 4%, but that may partly be due to its limited geographical coverage. It can only be received in around 70% of U.K. TV households, whereas the other channels have coverage of almost 100%.

8.2.2 A struggling cable industry

The cable-TV industry, during the 1980s, was promoted by the government as an opportunity to provide high-bandwidth interactive services

Table 8.1
Audience Shares in U.K. Television, 1997–1998 (in %)*

	March 1997	March 1998
BBC1	31.6	29.7
BBC2	11.8	10.3
ITV	34.0	32.6
Channel 4/S4C	10.7	10.5
Channel 5	—	3.9
Others	11.7	13.0

*Includes terrestrial, cable, and satellite viewing.
Source: European Television Analyst, Vol. 8, 1998, p. 9.

through entertainment-financed television. The introduction of cable TV was perceived not as a matter of broadcast policy, but of information technology policy. Government policy for the cable-TV industry is discussed in more detail in Section 8.6.3.1.

8.2.2.1 High investments and few customers

The 1984 Cable and Broadcasting Act allowed cable-TV franchises to provide both telecommunications and television services. This, however, excluded voice telephony. Only since 1991 have cable companies been able to offer telephony services. To attract investment, cable TV was not subject to content regulation (e.g., the 14% quota on foreign programs). Since 1990, the sector has been open to foreign investment. Today, the majority of cable-TV franchises in the United Kingdom are held by subsidiaries of North American telecommunications and cable-TV companies. These firms invested heavily to gain experience in providing both telecommunications and TV services—not allowed in their home markets at the time. Over the time period 1990–1997, around £5.4 billion has been spent on building the U.K. cable network to cover about 40% of U.K. homes.[1] During that time, the number of cable homes passed has risen from 0.5 to 8.4 million. By 2000, 12.8 million households had been passed. To protect their investment, the dominant telecommunications operators, British Telecom (BT) and Mercury, were excluded from the carriage of TV services over their telecommunications networks until 1998, and from ownership of such entertainment services until 2001 [1].

From 1994 to 1998, cable-TV penetration (connected homes in relation to passed homes) was static at just over 20%, while churn rates and capital investment have been high. By 1998, only around 2.5 million homes (11%) used cable TV. The prime reason for this poor development is the cable operators' dependence on content from BSkyB, and their inability to engage in effective price competition with this dominant operator.

8.2.2.2 Cable operators' dependence on BSkyB

In the early years of cable TV, the cooperation of the cable companies with BSkyB proved mutually beneficial. Cable operators received programming, and BSkyB acquired a customer base. The situation changed in 1992 after BSkyB acquired the exclusive live rights for Premier League

---

1   £1 is equivalent to approximately $1.5.

Soccer for a period of five years. Similar deals were then struck for rugby, golf, and cricket. With this sporting monopoly, which proved the driver for pay TV, BSkyB gained a dominant position over the cable companies.

The cable companies are in the unfortunate position of, on the one hand, depending on BSkyB premium content and, on the other hand, having to compete with it. Several times the cable companies complained to the Office of Fair Trading (OFT), arguing that the prices charged by BSkyB for access to its programming and the obligation to distribute BSkyB programming to all of their cable customers was anticompetitive. BSkyB charges cable operators £17 to carry the basic package plus one premium channel. Satellite subscribers pay £20.99 for the same package. This means that in order for cable operators to make a profit, they have to charge consumers higher fees than BSkyB itself. Although cable operators can justify higher prices by the provision of the decoder box free of charge, the prices of decoders as well as satellite dishes have been coming down dramatically over the last few years, leaving cable tariffs looking expensive.

The high charges by BSkyB also mean that cable has little revenue to invest in program origination. Contractual provisions mean that BSkyB can arbitrarily determine prices, force the cable companies to take new channels, and prevent them from unbundling channel packages.

8.2.2.3 Channel bundling distorts competition

An important and much discussed step in the monopoly exploitation of BSkyB is the bundling of channels. Channel bundling can be justified as it leads to economies of scale on the transmission level because the cost of adding a new channel to a bouquet of existing ones is relatively low. Similarly, once the distribution infrastructure is in place, the cost of connecting another household (subscriber) is minimal. In this sense, channel bundling can lead to more subscriptions as well as higher revenues. This can also translate into lower prices and a better service for consumers.

However, channel bundling can also lead to significant distortions of competition if it prevents market entry and, thus, limits the overall service offering and increases prices. There are four types of channel bundling:

- ▸ Buy-through: The subscriber of a premium (sports or movie) channel is forced to also subscribe to (and, thus, pay for) a usually less attractive basic package.

▸ Discounts: Subscribers receive additional premium channels at less than the average costs per subscriber.

▸ Free bonus channels: Subscribers receive additional channels at no extra charge.

▸ Minimum carriage requirements: The distributor (cable operator) has to distribute a channel to a fixed minimum percentage of subscribers, or pay higher prices.

BSkyB applies all four types of bundling. In particular the latter two are seen as distorting competition. Cable operators gain financial advantages if they transmit all BSkyB channels. If a particular channel is refused, the price for the others is higher. Furthermore, BSkyB provides financial incentives to market, in particular, the premium channels. Similarly, the cable operators are financially "punished" if they accept independent channels that provide competition to BSkyB. They are furthermore obliged to transmit all BSkyB channels to all connected households, and they have to adopt the same bundling strategy, allowing subscription of premium channels only in combination with a basic package.

The bundling strategy of BSkyB has led to the effect that cable operators are discouraged from offering alternative premium content and from increasing the number of basic channels in their offers. BSkyB has thus secured its monopoly position in premium content and retains control over the competition in basic programming.

The cable operators have lodged various appeals with the OFT in 1995 and 1996 on the grounds of restraint of trade. These proved almost entirely unsuccessful [2]. Since November 1996 the ITC has been looking into the issue of channel bundling and has since improved the situation for the cable operators (see Section 8.6.4.3).

8.2.3 A monopolized satellite sector

The U.K satellite-TV market is entirely in the hands of BSkyB. News Corporation holds 40% in the company, an amount that has been enough to ensure effective control. BSkyB began its life as Sky in 1979 with four free channels, mainly broadcasting to the European continent via cable networks. Its first movie channel was encrypted a year later. In 1986, the U.K. government licensed a commercial consortium, British Satellite Broadcasting (BSB). This group encountered technical difficulties in meeting the government's high technical specifications and as a result was six months later than Sky in launching a domestic DTH satellite service. With Sky being the first mover into the U.K. DTH market with a proprietary dish standard, and with consumer reluctance to buy more than one dish for competing content, BSB failed, and the two companies merged into BSkyB (see also Section 8.6.3.2).

Satellite TV only really began to take off, however, when BSkyB began to buy up exclusive sports rights. In September 1992, with the acquisition of monopoly rights to Premier League Soccer, the sports channel moved to encryption. One year later, its basic package followed with the introduction of 15 new channels. A pay-per-view system for premium sports events was introduced in 1996.

As noted previously, in the early years cable TV was more important than dish ownership for the distribution of BSkyB's programming. In 1989, BSkyB claimed 58,000 dishes receiving services direct from the Astra satellite, compared to 200,000 U.K. homes with cable TV. By 1997, of the 6 million homes reached by BSkyB, only 2 million were connected to cable TV. Only recently, cable has begun to catch up again. In August 1998, Sky had around 3.6 million DTH satellite subscribers.

The encryption system gave BSkyB a de facto monopoly on connection for all would-be satellite programmers for the U.K. market. In addition, because it broadcasts from nondomestic satellites (Astra satellites are operated by the Luxembourg-based SES), it is exempt from U.K. content regulation and, unlike the commercial TV companies, has to pay no broadcast license fee (around £350 million per annum). Moreover, despite News Corp.'s dominance in the newspaper industry, the Thatcher government also exempted it from U.K. rules on cross-media ownership [3].

In summary, it can be noted that there is potential for multichannel TV in the United Kingdom as most households still receive only a few channels. The multichannel pay-TV sector is strongly dominated by BSkyB. It is against this background that the strategies and regulatory approaches for digital TV are presented.

8.3 The United Kingdom pioneers DTT

The United Kingdom is the first country in the world where the government has made available terrestrial spectrum for the provision of digital services. Six blocks of frequency—so-called multiplexes—have been provided with 8 MHz of bandwidth assigned to each of them. They are located in so-called taboo frequencies, which cannot be used for analog transmission due to interference problems but can be used for digital transmission alongside existing analog services. Each multiplex can carry about six channels. Under the 1996 Broadcasting Act, existing terrestrial broadcasters have been allocated three multiplexes for the provision of DTT services to be launched in 1998. The other three multiplexes were offered in a bidding process for broadcasters to develop new services. The allocation and plans for the usage of the multiplexes are discussed in Sections 8.3.1 and 8.3.2.

8.3.1 Existing broadcasters

The BBC has sole use of one multiplex (Mux-BBC), ITV, and Channel 4 of another (Mux ITV/Ch4). Channel 5 and the Welsh fourth channel S4C will use half of a third multiplex. The other half of that multiplex, labeled Mux-A, was available for commercial bidders. S4C emerged as the only bidder via a new company S4C Digital Networks (SDN) and was awarded the license by the Independent Television Commission (ITC).

8.3.1.1 The BBC's digital offering

For digital transmission, the BBC not only extended its range of existing free-to-air channels, but also started to provide pay-TV channels.

Free-to-air channels   The BBC started digital transmissions in September 1998. In addition to simulcasting BBC1 and BBC2 in digital format, a range of new value-added channels and services were introduced. These include the following:

> ‣ BBC News 24: A round-the-clock news channel also available on analog satellite and cable TV since 1998;

> ‣ BBC Choice: A thematic channel complementing BBC1 and BBC2 content;

> ‣ BBC Digital Text: An expanded teletext service including audio and video as well as BBC Web content;

> ‣ BBC Parliament (to be launched during 1999): U.K. government coverage;

▸ BBC Knowledge (to be launched during 1999): Education and information programming from BBC Education, Open University, and repeats from BBC1 and BBC2.

Under public service obligations, all these services are available on a free-to-air basis for digital-terrestrial, digital-satellite, and digital-cable TV.

For the BBC, the transition to digital involves a difficult balancing act with regard to the license fee. The license fee in the United Kingdom is £101 (as of April 1999). The question arises of how services only available in digital homes can be funded out of a license fee paid by millions of nondigital households. The BBC review panel has looked into the possibility of charging a digital supplement to those households that have converted. A similar approach has been adopted in Japan, where NHK's satellite viewers pay more than the terrestrial viewers (see Section 9.2). Such a decision would create fairness for license fee payers but would not be welcomed by manufacturers, retailers, and other broadcasters aiming for a fast transition to digital, since it would create an extra cost to consumers and form a disincentive to switch.

Pay-TV offerings   In addition to the extension of the public services, the BBC is also offering pay-TV services. In March 1997, the BBC and Flextech, a fast-growing cable and satellite programming company, agreed to form the joint venture, UKTV, with the objective of launching eight new channels for the U.K. digital-TV market. Flextech will provide the risk capital in return for access to the BBC programming library and programming expertise. Moreover, the company hopes that the perceived BBC quality will be a crucial factor in differentiation. The advantage for the BBC is that it will be able to supplement the license fee and invest earnings from the new subscription channels into programming. Furthermore, the BBC has an interest in supporting digital pay TV: Digital decoders will only be offered at subsidized prices in conjunction with a pay-TV subscription; few viewers will be prepared to pay the full price (around £400) for a digital box only to watch the free channels. Hence, the BBC's involvement in digital pay TV helps to promote its public service channels.

In addition to the existing Flextech channels, U.K. Gold, Living, Challenge TV, Bravo, Trouble, U.K. Play, and U.K. Arena, some new channels were developed. Those are described in Table 8.2.

These channels are meant to be shown on all digital pay-TV platforms. They are not part of the BBC's public broadcast services and are

Table 8.2
New BBC-Flextech Channels for Pay TV

Channel	Content
U.K. Horizons	Factual programming
U.K. Gold 2	Comedy and drama
U.K. Style	Leisure and lifestyle
U.K. Drama	TV dramas
Screenshop	Home shopping
TVTravelshop	Travel shopping

therefore not available on the BBC multiplex. The BBC-Flextech channels contain mostly thematic content; hence their channels typically form part of basic multichannel packages. For example, the BBC-Flextech joint venture supplies six channels (U.K. Gold, U.K. Gold 2, U.K. Style, U.K. Horizons, U.K. Drama, and U.K. Play) to ONdigital as part of a basic package.

The ITC ruling on the abolition of minimum carriage requirements and channel bundling (see Section 8.6.4.3) forms a severe setback for the company, as the financing of channels with small audience shares often requires selling those as part of a larger package. The new ITC rulings, which took effect in January 2000, may result in some channels being discontinued. Already, plans for further channels, including a TV equivalent to Radio 1, a sports channel, and an educational channel, were abandoned.

8.3.1.2   ITV and Channel 4

ITV and Channel 4 have formed a new company, Digital 3 and 4, to manage their joint multiplex. They have modest plans since the largest ITV companies, Carlton and Granada, focus on their pay-TV venture, ONdigital (see Section 8.4.2), and Channel 4 intends to primarily exploit its film rights and archives.

ITV simulcasts its 16 regional programs terrestrially. In addition to this, a new channel, ITV2, was launched. This will initially broadcast 85 hours a week with a youth-oriented and alternative slant, without regional variations.

Channel 4's existing analog channel is also simulcast in digital. A second DTT channel, Film Four, was launched as a pay-TV channel. It forms a variation of the analog Film Four channel, a premium movie channel showing the films produced by Channel 4. It will also broadcast films that have not received theatrical release in the United Kingdom. Initially, this channel was exclusive to ONdigital viewers, but it is now also available to cable and satellite viewers. Since it is expected to have a smaller market, it has adopted a fairly aggressive pricing point of £6 per month.

8.3.1.3　SDN

SDN is a joint venture of S4C, United News & Media (UN&M), and NTL with each holding a 33% stake. NTL and UN&M were both part of the DTN joint venture that failed in the bid for the commercial multiplexes (see Section 8.3.2.2). SDN offers simulcasts of the terrestrial Channel 5, S4C (the Welsh version of Channel 4), and some Gaelic programs in Scotland. Channel 5, so far, does not plan to produce any new programming.

SDN initially held back plans for its channel lineup. In February 1999, it formed a joint venture with ONdigital to launch a pay-per-view service in autumn 1999. Movies sports, concerts, and events are offered on five channels of the SDN multiplex. This service competes directly with cable and satellite pay-per-view offerings. The remaining capacity on SDN's multiplex is used for interactive services from NTL, launched in July 1999 (see Section 8.5.2.3).

8.3.2　The bidders for the commercial multiplexes

The remaining three multiplexes (Muxes B, C, and D) were made available for commercial broadcasters. Successful bidders are granted 12-year licenses. It is worth noting that licenses for multiplex operation are separate from licenses for program provision. In the initial period, there is no charge for the multiplex license, providing an incentive to launch DTT services. The bidders were judged on the quality of their plans and the likelihood of a successful launch. The ITC's criteria for judging the license applications were as follows:

‣ Coverage of U.K. population;

‣ Speed of rollout of service;

‣ Ability to establish and maintain services;

‣ Appeal of program services to a variety of tastes and interests;

‣ Plans to promote and assist acquisition of decoders by viewers;

‣ The plan drawn up by applicants to ensure fair and effective competition in their dealings with providers of programming and additional services.

By the deadline of January 31, 1997, two consortia, British Digital Broadcasting (BDB)—later renamed Ondigital, then in June 2001 renamed ITVDigital[2]—and Digital Television Network (DTN), had entered bids for each of the multiplexes, although both expressed the hope that they would win all three. Section 8.3.2.1 examines the proposals of the bidders and the outcome of the license award.

8.3.2.1 The ONdigital consortium (formerly BDB)

The three shareholders in the original consortium, Carlton Communications, Granada Group, and BSkyB, agreed that, if all multiplexes were awarded, they would provide a peak funding requirement of £300 million. The planned services are summarized in Table 8.3. ONdigital would also offer pay-per-view services and an EPG. The consortium claimed it would actively seek arrangements with other multiplex operators to adopt a set-top box capable of receiving all DTT services plus digital satellite via an add-on module that should cost less than £50.

In general, it was clear that ONdigital would be showing a range of satellite channels and new channels similar to the existing satellite channels. There is very little original programming. It would thus not be selling a new service but creating a new distribution system (i.e., digital terrestrial) for existing multichannel offerings. This could open a large new market segment since many potential customers have resisted buying satellite or cable due to a perception that it provides "working-class" entertainment.

Although the BBC channels were expected to have strong appeal with the public, the credibility of the consortium was based mainly on the presence of BSkyB and its provision of film and sports premium channels. These have been the drivers of (analog) cable and satellite and are equally expected to be the drivers for DTT.

8.3.2.2 DTN

The other bidder, DTN, was a wholly owned subsidiary of International CableTel, a U.S. cable company that owns U.K. franchises as well as the

---

2. For simplicity's sake, the name ONdigital will be used throughout the text.

Table 8.3
Channel Services Proposed by ONdigital

Basic Subscription Package	Services
Sky One	BSkyB's existing satellite entertainment channel
Carlton Select	"Enhanced" version of existing cable channel, plus live European football
Carlton Entertainment	New channel with strong emphasis on British productions
Carlton Films	Old British films and U.S. movies
Public Eye	Sky News in morning and early afternoon; dramas, documentaries, and law-and-order films
Granada Plus	Existing satellite entertainment channel
Granada Good Life	Existing satellite lifestyle channel
Granada TV Shopping	Teleshopping
Granada Sports Club	News, opinion, debate, and entertainment on sports; also a late-night music channel
U.K. Style	Lifestyle programming
U.K. Showcase	Popular BBC1 and BBC2 programming
U.K. Horizon	Documentaries, with morning children's programming
U.K. One TV	Pop music
Premium Channels	
Sky Movies	Movies
The Movie Channel	Movies
Sky Sports	Sports

Source: Broadcast (February 7, 1997), p. 20; New Media Markets (February 6, 1997), p. 3.

transmission and cable company NTL. In its bid, DTN offered an altogether different package, placing much emphasis on special interest services and interactive content. Proposed channels are listed in Table 8.4.

DTN would also offer a local information service, Metro TV, the pay-per-view movie service Digital Box Office, an EPG, and up to 20 data services. Cheap telephony and subsidized decoders were also part of the package.

The offer contained more original programming but few general-interest channels and little premium content. A further factor unfavorable to DTN was that it could not gain support from the BBC. This issue was important because each bidder aimed to win the new channels planned by the BBC-Flextech joint venture as part of its channel line-up.

Table 8.4

Channel Services Proposed by DTN

Original Channels	Services
Money Channel	Finance for general audience
Knowledge Network	Education
British Sports Channel	Twenty-four-hour sports, including local and minority events
Specialist Channels	
ITN Living History	Old news repackaged as history
Animal Planet	Wildlife
Travel	Travel
The Box	Interactive music channel
International Channels	
Turner Classic Movies	Old Hollywood films
Cartoon Network	Cartoons for children with evening programming for adults
MGM Gold	Entertainment
The Movie Experience	Films

Source: Broadcast (February 7, 1997), p. 20; New Media Markets (February 6, 1997), p. 3.

8.3.2.3 License award to ONdigital

In June 1997, the ITC chose ONdigital on the condition that BSkyB withdraw its 33.3% stake in the venture. The other two members of the consortium, Carlton and Granada, agreed to buy BSkyB's share at £75 million to become 50:50 investors. The ITC also emphasized that BSkyB should still remain the primary program supplier to ONdigital. Considering that the ONdigital business plan suggested that BSkyB would generate more revenue from program supply (up to £1 billion over five years) than from shareholding, the fulfillment of this condition did not form a major disadvantage for any of the ONdigital companies.

The formal license for ONdigital's service license was granted on December 23, 1997. It had been held up by discussions with the EU competition authorities following a formal complaint made by DTN. With regard to competition concerns the ONdigital license stipulates the following:

▸ ONdigital should not be prevented from competing with satellite broadcaster BSkyB by Granada's 11% stake in BSkyB; in response to this, Granada sold this stake in October 1998.

▸ Program supply contracts have to be limited to five years (ONdigital had previously agreed to a seven-year deal with BSkyB).

▸ ONdigital is obliged to support open technical standards for integrated DTT sets.

The license further obliged ONdigital to start services within one year from the date of the award. ONdigital commenced transmissions in November 1998.

The license award was criticized by many industry experts and, in particular, by OFTEL, on the grounds that BSkyB's dominant position in the television market was maintained. From a viewer perspective, the ITC decision is justified, because it is undisputed that BSkyB holds more attractive content rights for U.K. pay TV. By forcing BSkyB to give up its partnership in the consortium, the ITC has effectively created a competitor to BSkyB, although ONdigital is dependent on BSkyB's premium content. The fierce competition that has developed between the two digital operators (see Section 8.4) supports this argument.

8.4   Business strategies of the digital pay-TV competitors

This section analyzes the launch strategies of BSkyB and ONdigital, who were first to offer digital pay-TV services in the United Kingdom.

8.4.1   BSkyB: Defending its dominant position

BSkyB has an established position in analog pay-TV services. The launch of digital terrestrial pay TV means that this position could now be threatened.   .

8.4.1.1   An incumbent in pay TV

BSkyB is one of the most successful European audiovisual firms with a turnover of £1.8 billion for the financial year 1999–2000. BSkyB occupies a pivotal position in the U.K. pay-TV market. It controls the analog CA system to satellite TV, provides a number of wholly owned or joint-venture basic channels, and, through its purchase of key sports and movie contracts over a relatively long term, controls the main premium channels in the United Kingdom—on both cable and satellite. BSkyB is the only operator in the United Kingdom with both a significant gateway control and programming assets. The BSkyB strategy to provide premium sports content experienced a setback when the proposed take-over of football club Manchester United was blocked by the government in March 1999. The deal was regarded as increasing BSkyB's market power as a sports channel provider and its chances of winning exclusive football deals.

The launch of digital terrestrial pay-TV services presents a serious challenge to the BSkyB monopoly. Average U.K. viewing time is 3.5 hours per day, and this figure is unlikely to increase significantly. Hence, the company now has to compete for pay-TV viewers. BSkyB's strategic moves are discussed in Sections 8.4.1.2 through 8.4.1.4.

8.4.1.2 SkyDigital: The launch offer

The launch of the digital satellite service SkyDigital had been delayed repeatedly. This was most likely due to two reasons. First, the uncertainty over the regulation of digital pay-TV services, especially CA (see Section 8.6.4.2), implied a significant risk. Second, BSkyB enjoyed healthy profits in analog pay TV; hence there was little incentive to launch digital services. Once CA regulations had been passed in the spring of 1997, BSkyB still delayed the launch of digital. This is due to the fact that CA is strictly regulated for digital services, whereas analog pay TV remains unregulated. BSkyB finally launched digital services shortly before ONdigital's services commenced so as to preempt the competitor.

BSkyB launched SkyDigital in October 1998, one month before the launch of ONdigital. In May 1997, BSkyB had ordered 1 million digital satellite decoders, worth more than £300 million. The four companies producing the boxes are Panasonic, Amstrad, Pace, and an alliance of Grundig and Hyundai.

Although BSkyB's head start was small (one month), BSkyB was determined to exploit its first-mover advantage and followed an aggressive and costly strategy to quickly win subscribers. BSkyB initially offered free installation of satellite equipment, costing it around £80 per subscriber, as well as a finance deal allowing consumers to purchase the required equipment (dish, decoder, and remote control) in installments of £5.99 per month. At full price, the equipment costs £200 to new subscribers and £160 to existing customers. The decoders are subsidized by

British Interactive Broadcasting (BIB), a joint venture including BSkyB and BT that is to offer interactive services on the SkyDigital platform (see Section 8.4.1.4). To raise consumer awareness, the launch of SkyDigital was accompanied by an advertising campaign with a budget of £56 million for the first year.

BSkyB has capacity to deliver around 200 digital channels via satellite. At present, it offers 140 channels. Included in these are simulcast versions of all existing analog channels in the Sky MultiChannel Package (including 18-to-20-hour-a-day versions of services currently only broadcast for a few hours a day), digital versions of all Sky's existing premium channels (transmitting four to five differently scheduled versions of Sky Movies), and a digital 40+ channel NVoD version of Sky's pay-per-view service Sky Box Office. This four-channel analog service, the first U.K. pay-per-view movie offering, started in December 1997. Each of the four channels on Sky Box Office screens one movie a week continuously during prime time (four screenings a night) at £2.99 per movie.

The SkyDigital program offering consists of a range of packages that can be subscribed to in several combinations. Six basic packages are available at prices between £7 and £12 per month. SkyValuePack is the least costly option (£7) comprising two entertainment, two news, and one shopping channel, plus 11 radio channels. Sky Family is the full basic package at £12 per month. It offers 54 thematic channels including channels from other providers (e.g., MTV, Discovery, Granada, or U.K. Gold). This offering also includes 13 free-to-air channels and 30 radio channels.

On top of a basic package, viewers can chose among a range of premium packages depending on the number of channels desired. Most importantly, SkyMoviesWorld, at £27 per month includes SkyFamily plus three premium movie channels from Sky, which are shown on up to five channels each at staggered starting times. Also included is the Disney Channel and access to Sky Box Office, a pay-per-view service for movies and events. SkySportsWorld is the equivalent for the sports viewer: Also at £27 per month, it includes SkyFamily plus four premium sports channels from Sky. The complete range, SkyWorld, comprising movie and sports packages, costs £32 per month.

Four separate premium channels (Disney Channel, MUTV, Artsworld, and Music Choice Extra) are available at £5 or £6 each.

BSkyB claims that practically all subscribers (98%) choose the full basic package "family" rather than one of the smaller basic bouquets. Pay-per-view appears very attractive to digital viewers: It is said to be 10 times more popular than it is in analog due to the larger choice available.

By January 1999, BSkyB had around 350,000 digital subscribers—only four months after launch. This makes it the fastest ever take-up of digital TV: So far, the most successful digital launch was DirecTV in the United States, which achieved 200,000 subscriptions during its first four months (see Section 10.1.3.2). Of the registered digital subscribers, 120,000 were new customers (34%) whereas 230,000 were existing analog subscribers. By May 1999, total subscribers had risen to 551,000, and the share of new subscribers had risen to 39%.

8.4.1.3 Additional competitive strategies

A closer look at subscriber developments reveals a slightly different picture. In the first quarter of 1999, more analog subscribers were lost (108,000) than digital subscribers had been gained (95,000); hence, total BSkyB subscriptions had actually fallen. This can be seen as an effect of the new competition from DTT. To stop this trend, BSkyB started several initiatives to make its offer more attractive.

Most importantly, since June 1999, BSkyB has been giving away set-top boxes at no charge with any one-year subscription to SkyDigital. Viewers only pay a £40 fee for the installation of the box. It has to be noted though that customers have to wait around seven weeks before the box is installed. This forms an obstacle to impulse purchases. ONdigital boxes, in contrast, do not require installation but can be used immediately. To push subscriptions, BSkyB extended the distribution channels for set-top boxes: While they were distributed mainly by electronics retailers, they are now also available in supermarkets and at toy retailer Toys R Us.

The free box offer can be expected to be unpopular with customers who have already paid £200 for a set-top box. To appease these customers, BSkyB announced that their subscription fees would be fixed until 2001, whereas those for new subscribers and existing analog subscribers would rise. From July 1999, the basic packages increased by £2 to £3 per month and the premium bouquets by £2. The minimum basic package, however, stayed at £7.

With regard to pending competition from cable operators, BSkyB (in connection with the News Corp. telecommunications operator Broadsystems) offers reduced phone call pricing to SkyDigital subscribers. Costs on phone calls are guaranteed to be 40% lower than BT rates.

The range of incentives offered to customers to attract them to SkyDigital has required large investments. For example, acquisition costs for each new subscriber are estimated to be £155. These are made to ensure

BSkyB's future dominance in the U.K. pay-TV market. The fast uptake of digital pay TV in the U.K. supports BSkyB's assumption that this will quickly become a large market with significant revenues to be earned. The company appears to be convinced that it will be able to recoup its initial investments in the long run.

So far, these assumptions have proved correct. The free box offer gave a boost to subscriber numbers, reaching around one million customers in August 1999, two months ahead of schedule. By mid-2000, Sky-Digital had 3.6 million subscribers. The number of analog satellite customers had decreased to 930,000. With this favorable development, BSkyB now plans to have all its analog viewers converted by June 2001, when the analog service is to be switched off—18 months earlier than originally planned.

8.4.1.4 Interactive services from "Open"

In addition to the large program offering, the BSkyB platform provides interactive services through the BIB joint venture founded in mid-1997. The main shareholders in the BIB joint venture were originally BSkyB and BT, with 32.5% each. The bank HSBC (formerly Midland Bank) held 20% and Matsushita the remaining 15%. In November 1998, BIB chose "Open" as the brand name for its interactive services.

The forming of BIB had been under review from national and European regulatory bodies since the cooperation involved large communications players such as BSkyB and BT, which are already dominant players in their respective markets. In late 1998, the venture was cleared by the European Competition Authorities after BSkyB and BT had given concessions. The companies pledged that third parties would have access from launch to the BIB-subsidized decoders and the software needed to create and run interactive services. In practice, this will mean allowing third parties to use the proprietary CA and API software in the BSkyB decoder, which checks whether downloaded applications are authorized and whether the viewer has a valid smart card. Furthermore, BIB guarantees that buying its decoder does not obligate subscribers to take any pay-TV services.

BSkyB's interest in the cooperative effort stemmed originally from the possibility of retailing digital satellite decoders at the subsidized price of £200 rather than the £400 they would initially cost. The BIB joint venture would provide £265 million in subsidies for the set-top boxes. Over a 10-year period, BIB claims to be prepared to invest up to £700 million in subsidies. The venture is projected to break even after five years. The positive response to the Open services resulted in BSkyB assuming majority control of Open in July 2000. The broadcaster now holds 80.1%, with BT holding the rest.

BT's ambitions in the venture include the potential provision of services such as VoD and educational programs over its telephone lines. Current regulations prevent it from broadcasting entertainment services on a national basis, but it can transmit programs on an individual basis. Furthermore, since May 1998, BT is allowed to offer broadcast services to households that are not in a cable franchise area.

Decoder technology   To receive the Open interactive services, the BSkyB decoders need to be upgraded. This is done by downloading the API. It has to be noted that Open initially only provides a traditional narrowband service with some broadband video ("video-enhanced") but no full interactivity.

The decoder box is equipped with BSkyB's proprietary CA system NDS. Open, however, announced that it would not use the OpenTV API, the one selected by BSkyB, as this would restrict access to SkyDigital subscribers only. Open aims to make its services compatible with a wider range of technologies including the Internet. This strategy is pursued to enable the company to establish itself in other markets such as DTT or digital cable where OpenTV will not be used. In order for interactive services to become profitable, it is regarded as crucial to have access to a large market. Being confined to BSkyB's digital subscribers might not be sufficient. It must gain carriage on other platforms so that it can offer a large customer base—an important factor when it comes to attracting service providers (e.g., retailers) to use its platform.

Service offerings   The timetable for the launch of Open services has continually been postponed. Full launch of the Open service commenced in October 1999. It is currently offered to SkyDigital subscribers only. For a one-off connection fee of £40, customers get access to the following:

> ‣ Free e-mail (BT's Talk 21), using an optional infrared keyboard;

> ‣ More than 20 shopping opportunities (from supermarket Great Universal Stores, Iceland, Woolworth's, Somerfield, catalog retailer Argos, electronics retailer Dixons, ticketing operator First Call, book and CD retailer WH Smith, travel operator Going Places,

football club Manchester United, listings services Yellow Pages, and music/sports/film e-commerce from PA News Media);

‣ Six to seven interactive information services (learning, weather, news, sports, and entertainment information);

‣ Five to six on-line games from European developer Visionik;

‣ Financial services from HSBC and Abbey National.

Backed by £375 million of investments, Open hopes to reach 15 million homes within five years. There are no charges for dialing into Open services, and no monthly subscription fee, though charges may be levied by some of the individual services and on-line betting appears to be highly popular.

Despite losses of around £300 million in relation to a turnover of £22 million over the first nine months, Open assesses its venture a success. The company claims that, as of mid-2000, more than 1.6 million subscribers use Open services at least once a week and 10% of users have conducted a purchase via the TV. A total of 750,000 users registered for the e-mail service, and on-line betting appears to be highly popular.

With BSkyB in control, Open services are now increasingly being integrated with SkyDigital channels. In June 2000, SkyNews Active launched as a complement to the Sky News channel. Sky News Active offers news on demand and extended coverage of live events in text and video format. SkySports Active, launched in August 2000, complements the SkySports Extra channel, allowing interactive applications such as sports statistics, slow motion repetition, highlights, and viewing from different camera angles. Open plans further program-related interactive services such as direct ordering of sports merchandise.

It has to be noted that Open pursues a "walled garden" strategy, allowing access only to a range of selected information and commerce offers, rather than access to the full Internet. This strategy is suitable in the early stages, to familiarize TV viewers with interactive services. However, it restricts consumer choice and can therefore only be a medium-term strategy. ONdigital, in contrast, opted to offer access, not to all, but to a large number of Internet sites, and it is expected that Open will follow suit when introducing the next generation of set-top boxes.

8.4.2   ONdigital: A successful second mover

ONdigital, the joint venture of Carlton and Granada, launched digital terrestrial pay TV in November 1998—only one month after the SkyDigital launch. From the start, it aimed to differentiate its offering from that of BSkyB. It quickly became clear that these two players would become fierce competitors.

8.4.2.1 Strategic positioning

Shortly after the DTT license award, ONdigital started to distinguish itself from BSkyB. Prior to the launch of digital services, it was expected that BSkyB and ONdigital would use compatible decoders so that viewers would be able to choose between offerings. This would mean that the two operators would compete on the basis of programs. As BSkyB can offer many more channels and owns most premium content, it was expected that ONdigital would eventually lose out in this competition.

In early 1998, however, ONdigital announced that it would set up its own network with Seca's MediaGuard CA system (used by Canal Plus), which is not compatible with BSkyB's system. ONdigital aimed to lock in subscribers and maintain its subscriber base once established. In response to this announcement, BSkyB threatened legal action against ONdigital, claiming that ONdigital had both a regulatory and contractual obligation to ensure decoder compatibility with BSkyB. An agreement to this effect is believed to have been forged at the time of BSkyB's enforced withdrawal from the ONdigital consortium, in June 1997. BSkyB failed to overturn ONdigital's decision.

The employment of an incompatible CA system can be seen as a "war declaration" in the digital-TV market. Both companies would have to engage in strong competition so as to quickly gain an installed base of subscribers and trigger positive feedback.

The decoders for ONdigital boxes were initially produced by Philips and Pace, later also by Grundig, Nokia, Sony, and Toshiba. It is estimated that decoders cost around £400 apiece but they were initially subsidized and retailed at £200 (i.e., the same price as SkyDigital boxes). The specifications of the ONdigital box design were agreed on by all DTT broadcasters. The key features are listed as follows:

- Two common interface sockets, which could be used to add on a decoder for digital cable or satellite;

- An embedded CA system (MediaGuard) but the possibility for a second to be added via the common interface;

‣ No API but use of the MHEG-5 software standard, which allows enhanced teletext services and EPGs.

ONdigital perceived the provision of idTVs as an important strategy move, so as to make use of the TV replacement cycle: Households typically replace their TV sets every 10 to 15 years. If digital sets are available at a cost similar to analog sets, it was seen as likely that consumers would acquire a digital set. Once acquired, the threshold to pay TV is low since they would only have to order a smart card to receive services.

Integrated digital-TV sets with the ONdigital CA system have been available since 1999, and prices are coming down rapidly. British electronics manufacturer Alba offers idTV sets (under its low-cost Bush brand) for as little as £500, whereas previously consumers were faced with a cost of approximately £1,000. These viewers may be interested in subscribing to ONdigital at a later date. The main potential customers for ONdigital would be those who are unwilling or unable to use a satellite dish and are not connected to a cable network. This is thought to apply to around 40% of U.K. households. ONdigital can serve the subset of those who are interested in pay-TV services.

However, as of mid-2000, only around 15,000 idTV sets had been sold in the United Kingdom. This illustrates the significance of equipment subsidization: Few consumers are prepared to invest in idTVs or set-top boxes.

8.4.2.2 The launch offer

ONdigital launched its service on November 15, 1998, backed by a £100 million advertising and marketing budget for the first year. The launch offer included the promise to provide free aerial replacement/installation for subscribers having trouble getting a signal within a good reception area.

ONdigital's programming strategy is based on the assumption that people do not want to watch more than six to ten channels and tend to devote two-thirds of their viewing time to channels with which they are familiar. ONdigital CEO Stephen Grabiner claimed that "our research shows most people don't want hundreds of channels. They associate it with poor quality. They want a few more channels of high quality, and that's what we're offering" [4]. Furthermore, ONdigital intends to differentiate itself from BSkyB's packaging and pricing strategy, which, it claims, created consumer resistance to large packages of channels.

ONdigital, in contrast, aims to offer smaller, flexible packages at attractive prices.

ONdigital was the first to introduce "pick 'n mix" channel packages. Subscribers can select any six from a choice of 18 basic channels at £10 per month or all 18 at £12. In combination with a basic package, a range of premium channels are available, including SkySports 1, 2, and 3; Sky-Premier; and MovieMax. Again, viewers can choose: One costs £11, two cost £15, and all five are offered at £18 per month (in addition to a basic package). Further single premium channels are the Manchester United football channel (£5) and the Channel 4 service Film Four (£6). A pay-per-view service with events (at varying prices) and movies (at £3) is also available. As pay-per-view has proved popular with BSkyB's digital viewers, in 1999, ONdigital launched an additional five-channel pay-per-view service, called ONrequest, together with SDN (see Section 8.3.1.3).

The main channel providers for the basic package are the partners Carlton and Granada, drawing on content libraries of 9,000 and 20,000 program hours, respectively, mixing archive programming with low-cost lifestyle programs. Granada is also offering the home-shopping service "Shop!" in conjunction with the retail company Littlewoods. This channel is provided free with any package. A new service, "First ONdigital," is a showcase service rather than a full channel and broadcasts special events and original commissioned programs ahead of any other transmission. A further source of programming is UKTV, the BBC-Flextech joint venture. Most of the premium channels are provided by BSkyB.

An important differentiation from SkyDigital is the exclusivity of ITV and ITV2 on DTT. The ITV channel is highly popular with viewers, gaining the highest viewing shares (33%) of all U.K. channels. ITV2, a new channel, is expected to attract viewers as it screens UEFA Champions League soccer as well as exclusive editions of popular ITV programs. BSkyB filed a complaint against the exclusivity of the two channels, but the ITC and the OFT decided in March 1999 that there were no grounds for requiring the channels to be supplied for digital satellite distribution.

8.4.2.3 Additional competitive strategies

To keep up with competitors, ONdigital is aiming to continuously improve its service. Three new channels (The Nursery Channel, Rapture, and Sky Sports 2) as well as digital teletext and interactive games were added from March 1999. Digital teletext has been made available through the downloading of an MHEG-5 API to decoders.

Free give-away of set-top boxes spurs subscriber take-up   The commercial DTT venture fared better than expected by many industry experts. Over the first five months, ONdigital gained around 110,000 subscribers. Growth was further increased when, in June 1999, ONdigital decided to give away set-top boxes to consumers who subscribe to an ONdigital package for at least one year. The set-top box give-away added 68,000 subscribers to ONdigital within one month, lifting total subscribers to 247,000 by the end of June 1999. The free-box offer was made in response to the move by SkyDigital, and with regard to pending competition from cable operators who announced launch of digital services from July 1999 and who traditionally give out decoder boxes for free (in fact, decoder rental is included in the subscription).

By December 2000, ONdigital had passed the 1-million-subscriber mark. The company now aims for 2 million subscribers by 2003. This is a remarkable success for the world's first digital terrestrial pay-TV service, which also has to compete with a very strong pay-TV incumbent, namely BSkyB.

Advanced interactive services   A range of new interactive services was introduced during 2000. In March 2000, ONmail, an e-mail service through a telephone connection was launched. Users pay a one-time connection fee of £30. After that, they only incur local phone call charges for sending and receiving e-mail.

ONdigital launched an Internet access service in September 2000. Again this is through a telephone line. Required are a small Internet box (produced by Netgem), an infrared keyboard, and a 56K modem. This equipment is rented at £5 per month. BT is the Internet service provider, charging 2 pence per on-line minute (1 pence off-peak). The Netgem box contains software that adapts Internet pages of ONnet's interactive service partners for TV viewing. This means that sites selected by ONdigital are easier to view and use, but consumers also have access to a large number of Internet sites. Selected sites are primarily shopping, banking, and travel services.

The EPG ONview represents the TV portal for DTT services. It forms the hub of entry, aims to provide easy navigation, and contains some commerce services and interactive advertisements.

ONdigital is also expanding program-related interactive services. Since February 2001, the ONsport service has provided program-related sports information and play-along games. ONdigital plans to introduce

Netlink, the possibility to link a program directly to a related Web site, while the ongoing program is still visible.

8.5 A new chance for the cable industry?

Cable operators have been late to offer digital services; however, they are considered to have good potential to improve their market position in the digital environment.

8.5.1 The advantages of digital cable

Digital transmission is widely regarded as a major opportunity for the stagnant cable industry to improve penetration rates. In accordance with the EU Directive on digital TV [5], cable-TV operators have a right to "transcontrol" (i.e., satellite as well as terrestrial programming has to be made available for cable transmission). This means that cable operators have the right to transmit BSkyB and ONdigital programming while administering the customer base themselves. As was noted in the analog case, problems may arise with regard to the price and conditions that BSkyB will try to determine.

An important competitive advantage for cable operators is to be able to offer smaller, more flexible packages (or even single channels) for subscription. So far, the cable operators have been limited by contractual obligations with BSkyB, including minimum carriage requirements and predetermined channel packages. Minimum carriage requirements and the bundling of channels were abolished in January 2000 (see Section 8.6.4.3), providing more freedom for cable operators to differentiate themselves from other digital platforms.

If DTT helps to expand the market for multichannel TV and induces a culture in which people are used to paying for services, this may assist the cable industry in the long run, since cable can, in principle, offer full interactivity and more capacity. The main barrier is the requirement to upgrade cable networks. To provide broadband interactive services, optical fiber infrastructure has to be laid all the way into the household. For this very expensive "last mile to the customer" several plans are being considered and tested at present. Furthermore, the cable companies have to improve their marketing and pricing strategies and provide better customer services to be competitive with DTT and digital satellite.

Potential competition may also arise from the lifting of the broadcast ban for BT. Since May 1998, BT has been allowed to offer broadcast services to the 17% of homes outside cable franchise areas. From 2002, the telecommunications company will be able to compete nationwide. Hence, cable operators have to establish themselves firmly in the market before that date, so that they will be able to effectively compete with this new entrant.

8.5.2 Digital cable services

This section looks at the strategies of the two remaining cable operators in the United Kingdom: NTL and Telewest.

8.5.2.1 Industry consolidation

Following a phase of strong consolidation during the 1990s, the U.K. cable industry had been reduced to three operators, Cable & Wireless Communications (CWC), Telewest Communications, and NTL. In 2000, NTL took over CWC, leaving only two competitors in the cable TV market.

Industry experts speculate that Telewest might join them, so as to provide a single, strong cable proposition in competition with digital satellite and terrestrial. Such a large company could seriously challenge BSkyB for sports and movie rights.

This speculation regarding the creation of a "super cable operator" was fueled by Microsoft, which already has a 5% stake in NTL, taking a 30% stake in Telewest in May 1999. Microsoft's objectives are to establish the WindowsCE API before a de facto standard emerges. For the same reason, Microsoft has expressed an interest in the German cable network (see Section 7.3.7). It is expected that the WindowsCE operating system will allow broadcasters and advertisers to provide interactive content at a fraction of the cost of BSkyB's Open service.

The creation of a single cable operator would need the agreement of regulators. However, in light of the dominant positions of BSkyB and BT, it is expected that regulators would be favorable to the idea of a strong cable competitor. BSkyB has dominated both the programming and the distribution of U.K. pay TV, and BT still has the majority of residential telephone customers.

8.5.2.2 Strategies of the cable operators

Prior to the merger announcement, the three cable operators had separate strategies for the introduction of digital. All three aimed to use General Instrument's DigiCipher CA system (also used by Echostar in the United States) in their set-top boxes, but this will be incompatible with those of BSkyB and of ONdigital. With regard to APIs, the cable operators were initially divided. Although all cable operators opted for Web-based (Java) Internet protocols as operating software in their decoders, CWC and Telewest chose Liberate software, whereas NTL selected PowerTV. In late 2000, following NTL's acquisition of CWC, the company announced it would also use Liberate software and ditch PowerTV. In terms of cable modems, NTL also follows the other two: It will adopt DOCSIS standard modems and phase out existing DAVIC cable modems. CWC was already using DOCSIS, as is Telewest. This means that all U.K. cable operators now employ the same technologies. This decision by NTL again mounted speculation that the company may bid to take over Telewest, leading to final consolidation of the U.K. cable market.

NTL's service offering   NTL and CWC together have access to around 10 million households in the United Kingdom and a subscriber base of around 2 million.

NTL has an advantage over the other cable operators because it also manages the transmission networks for the (analog) commercial terrestrial broadcasters in the United Kingdom and is a leader in DTT technology. Furthermore, NTL is a stakeholder in SDN; hence, it has access to digital cable and digital terrestrial platforms.

NTL adopted a different approach to the launch of digital services. Whereas most operators launch TV channel services first and then add interactive TV, NTL launched digital interactive services prior to digital channels. The company invested £25 million in a nationwide digital interactive TV service in April 1999, using the terrestrial multiplex of SDN and offering home shopping, news, sports, travel, and local information. To receive the services, NTL's TV Internet set-top box is required. The box uses a telephone return channel. Partners include Flextech, Tesco, Sporting Life, Ticket-Master, and Thomson Directories as well as the broadcasters ITN and BBC. The offer replicates some of the package that had been put together for the DTN bid, in which NTL was a partner. Those services were later also rolled out to digital cable.

NTL (and CWC) launched digital services in July 1999. In combination with cable TV, the company also offers phone line and dial-up, metered Internet services. Similar to its analog service, which reached high penetration rates (up to 37%), the company focuses on a few small packages rather than a big basic offering.

The basic package Digitalplus, at £12 per month, offers the five national free-to-air channels; the four BBC digital channels; ITV2; some bonus channels for sports, shopping, news, and travel; and access to the pay-per-view service FrontRow with movies at £3. The Digitalplus offer costs the same as a single standard phone line with BT; hence, this offer is attractive even for people who are not specifically interested in multi-channel TV. Moreover, the same option is available with unmetered Internet access at £17—cheaper than many Internet-only service providers.

Digital pay-TV services start with five thematic packages at £12 to £28 per month, depending on the number of packages chosen:

- Entertainment: Ten thematic channels for the whole family;
- Fun: Nine thematic channels for the whole family;
- Life: Nine lifestyle channels;
- Music: Nine music channels;
- World: Eleven news and documentary channels.

Viewers can change the number and combination of packages on a monthly basis.

On top of a basic pay-TV package, a range of premium channels can be selected individually:

- Seven sports and movie channels (mostly from Sky) are available at £15 to £23 per month.
- Six special-interest premium channels (including Disney Channel, FilmFour, two Asian-language channels) are available at £5 to £6 each.
- Three adult channels cost £7.50 each.

By the end of 2000, NTL had 531,000 digital cable-TV subscribers, up from 230,000 in mid-2000. This illustrates increasingly high growth rates. NTL was relatively late in launching digital cable service, but it is now rapidly gaining digital customers. The cable operators have encountered problems with the technology and with poor customer services, resulting in poor publicity. However, with its attractive combination of phone, Internet, and TV services, it can be expected that NTL will be more popular in the future.

Telewest's Active Digital service   Telewest is the United Kingdom's second largest cable operator with interests in 43 U.K. franchises and, following a recent merger with General Cable, a coverage of around 4.4 million households. The principal shareholders in Telewest were the U.S. cable companies MediaOne (30%) and TCI (22%). In May 1999, Microsoft acquired MediaOne's 30% stake.

Telewest had around 1 million analog cable-TV subscribers as well as 1.3 million residential telephone customers as of mid-1999. The company launched a digital service, Active Digital, in late 1999.

The "digital starter" package comprises 14 digital channels (including the 5 national free-to-air channels, 4 BBC digital channels, ITV2, ITN News, and 3 shopping channels) and a phone line for £9 per month. This offer also includes an e-mail service and access to the pay-per-view service FrontRow.

The "essential" package, at £15 per month, extends the starter package with 14 thematic channels. The "supreme" package comprises 70 thematic channels plus phone line at £22 per month.

On top of the thematic channel offerings, 14 premium channels can be chosen individually:

- Six Sky sports and movie channels are available at £11 to £20 per month, depending on the number of channels desired.

- The Disney Channel and FilmFour cost £6 each.

The Telewest digital cable packages are slightly cheaper than those offered by NTL. However, the choice of basic packages is smaller, and Telewest does not yet offer Internet access services.

The launch of Telewest's digital service was significantly hampered by delivery problems with the set-top boxes. That is brought forward as the reason why Telewest only gained 200,000 digital subscribers by mid 2000. Since then, however, take-up has still been slow and by the end of 2000, Telewest counted 315,000 subscribers, in contrast to the expected 500,000.

Cable still hampered, but has potential   Around half of all U.K. households (12.8 million) are passed by cable infrastructure, but still only 13% of those subscribe to cable-TV services (see Table 8.5). And only a fraction (13%) of those are digital-cable subscribers. The majority of cable operators' revenues is still generated from telephony services, which are used in nearly 5 million homes.

Table 8.5
Cable TV Connections in the United Kingdom (mid-2000)

Households	Connections (millions)
Passed	12.827
Total connected (phone and TV)	4.786
Connected to cable TV	3.347
NTL	2.166
Of which digital	0.23
Telewest	1.122
Of which digital	0.2
Other cable operators	0.59

Source: Media Perspektiven (October 2000), p. 442.

Technology problems and poor customer service have resulted in customer losses in favor of ONdigital and SkyDigital. Industry experts estimate that NTL and Telewest may have lost 50,000 to 100,000 customers to SkyDigital. However, it is expected that digital will play a larger role in the future, as cable operators can offer integrated phone, Internet, and TV services. Current subscriber growth of NTL, which has those offers in place, supports this argument.

8.6 A focus on regulation

The United Kingdom is at the forefront of digital-TV regulation. At the same time, the regulatory structure is very much influenced by past regulatory initiatives. This section explains the regulatory framework and illustrates how digital-TV regulation has been approached.

8.6.1 The regulatory system

The regulatory system in the United Kingdom is characterized by a variety of regulatory bodies, often with overlapping responsibilities. This section analyzes the situation and the resulting problems.

8.6.1.1 Telecommunications regulation

The task of devising the overall framework of telecommunications policy, within the Telecommunications Act of 1984, is the responsibility of the

British secretary of state. In addition, this post and the Director-General of Telecommunications [head of the Office for Telecommunications (OFTEL)] share certain generally specified duties such as protecting consumers, developing competition, and ensuring the supply of telecommunications services. The Director-General also has special responsibilities relating to the implementation of changes in licenses and other matters.

The licenses issued can be divided into two broad types: individual licenses, which are granted to public telecommunications operators (PTOs), and class licenses, which are granted to a specified class of persons (or more usually to all persons apart from certain specified exceptions, such as PTOs). The regulatory authority responsible for telecommunications is OFTEL. Prior to 1991, the only two national PTO licenses in existence had been awarded in respect of fixed networks to BT and Mercury. In March 1991, the U.K. government reviewed the BT/Mercury duopoly, within the white paper "Competition and Choice: Telecommunications Policy for the 1990s." Since then, the government has operated an "open-licensing" regime, whereby any applicants for a license to offer telecommunication services over fixed links within the United Kingdom will be equally considered. By 1996 more than 150 operators, including cable companies, had received licenses from OFTEL.

8.6.1.2 Broadcast regulation

In contrast to the regulation of telecommunications, which governs the conveyance (i.e., the delivery of entertainment services), broadcast regulation governs the provision of those services (i.e., the making and packaging of programs into channels). In broadcasting, Parliamentary Acts and Royal Charters lay down general obligations. Their detailed interpretation is the responsibility of several different regulatory bodies with specific responsibilities for particular aspects of the content of broadcasts. These include the ITC, the Broadcasting Standards Council, and the Broadcasting Complaints Commission (the latter two were merged in April 1997) as well as the Radio Authority.

Cable television is regulated under the Cable Authority, which became part of the ITC in 1990. A local delivery license, which is awarded through a competitive bidding process, confers exclusive rights to provide multichannel television over a specified service area for a period of 15 years. In practice, this means that each cable operator effectively holds a regional or local monopoly for cable TV. The license is technologically neutral (i.e., it does not specify the means of distribution).

Rights for terrestrial broadcasting are issued to the BBC under its charter and to other operators by the ITC, under the terms of an act of Parliament. There are currently three commercial, advertising-supported terrestrial channels (ITV, Channel 4, and Channel 5).

The Broadcasting Act providing for digital TV came into effect on October 1, 1996. Among other provisions, it translates the general principles of the 1995 EU Directive [6] into national legislation. It lays down the provision of six digital multiplexes for DTT and their distribution among broadcasters. This procedure is described in Section 8.3. The act also required existing terrestrial broadcasters to declare by October 15, 1996, their intention to take up their guaranteed places on the terrestrial digital multiplexes. All existing broadcasters did so. Furthermore, the 1996 act introduced some media ownership changes that took effect from November 1, 1996. Section 8.6.2 discusses these changes.

8.6.1.3 Problems with the current regulatory framework

The British system of regulation has evolved to become somewhat self-regulatory in nature, in contrast to the legalistic culture found, for example, in the United States. This is illustrated in the self-regulated nature of the BBC, which is under the control of a board of governors, appointed by the government of the day.

Another characteristic of British regulation is the tendency to create appointed bodies to oversee particular technologies, such as a designated Cable Authority, methods of financing (advertising versus license fees), or even responses to particular political concerns (e.g., pornography and violence). The 1990 Broadcasting Act rationalized the number of regulatory bodies to some extent by merging the Cable Authority into the ITC, which is now responsible for the private broadcasters on all distribution platforms. In 1996, the Broadcasting Standards Council and the Broadcasting Complaints Commission were also amalgamated.

A detailed 1995 study identified the following problems with the current regulatory framework [7].

▸ Duplication: Several firms are subject to multiple and overlapping regulation, not apparently guided by a clear and explicit organizing principle (e.g., the regulation of cable companies by the ITC as well as OFTEL).

▸ Uncertainty over the power of regulators: In some cases, the regulatory framework does not explicitly state which regulatory body, if any, is responsible [for example, VoD delivered over telephone lines through asymmetric digital subscriber line (ADSL) technology].

> Uncertainty arising from regulatory discretion: Regulatory agencies can decide how much or how little to disclose about their long-term intentions. Interactive effects make this process especially damaging for firms regulated by more than one regulatory body.

> Regulatory capture: Regulators are tempted to vie for power in the political market place by favoring their own regulatees (such criticisms have particularly been directed against broadcast regulators);

> Lack of accountability: Although regulators have a parliamentary mandate, arrangements for feedback to the parliament are weak.

> Nonequivalence of rules/tilted playing fields: Demonstrated by confused rules on cross-media ownership, arguments about line of business restrictions, and so forth.

The U.K. system is characterized by a variety of regulatory bodies with different and often overlapping responsibilities, similar to the situation found in Germany (see Section 7.4.1). Moreover, it creates competition between regulators and thus aggravates the implementation of effective regulation. To avoid such overlaps, the regulators have, for certain areas, created joint project teams where ITC, OFTEL, and OFT work together. Regulators now envisage eventually introducing a single regulatory body, called the Office for Communications (OFCOM). This has been suggested as a possibility in the government's green paper on "Regulative Communications" [8]. The ITC, however, is against such a radical overhaul of the regulatory structure. It reaffirms its commitment to close cooperation with the OFT and OFTEL in areas with overlapping responsibilities, but at the same time highlights the cultural aspects of broadcasting regulation and expresses its concerns that these would be lost within a single regulator. This protest from the ITC could possibly result from the organization's fear of losing power against the "stronger" OFTEL, but it certainly reflects a reluctance of the regulator to structural changes so as to achieve a higher efficiency.

It can be summarized that the United Kingdom faces the need for a fundamental reform of the existing regulatory system. This is in addition to regulatory challenges resulting from new digital markets, increasing liberalization, and converging ICT markets.

8.6.2   Media ownership regulation

After an extensive review the government concluded that there was a need to liberalize the ownership regulations introduced in the 1990 Broadcasting Act to reflect the needs and aspirations of the industry against a background of accelerating technological change. The Broadcasting Act of 1996 amended the provisions of the 1990 Act regarding restrictions on the holding of broadcasting licenses. Such measures did not constitute a complete deregulation. The government still regards the media as an industry that should not be subject to general competition law but needs additional regulation to guarantee plurality and diversity.

Media ownership regulation remains confined to the three most dominant sectors of the industry—namely TV, radio, and newspapers. The government does not perceive a case for extending media regulation to periodical publishing and "narrowcast" media such as on-line services, except to the extent that on-line video services are already regulated under Section 46 of the 1990 Act [9].

The 1996 Act permits a higher level of concentration in the TV market. A commercial broadcaster can now gain up to 15% of the national television audience. The medium of transmission (cable, satellite, terrestrial) as well as the number of channels it owns are not relevant. This new 15% audience share limit accommodates the considerable degree of consolidation of ITV companies that has occurred since 1990 and even allows scope for further concentration.

With regard to digital-TV multiplex services, the 1996 Act determines that no one person may hold more than three licenses to provide those services. A holder of three licenses is limited to a 20% interest in a fourth television multiplex license.

To control the provision of digital program services, the 1996 Act introduces a digital points scheme for television. The purpose of the scheme is to prevent a broadcaster from gaining a monopoly in the provision of digital-TV program services in the early stages of digital terrestrial TV. The points scheme is designed to stop an accumulation of interests in program services across the digital multiplexes. It does not apply where the services provided by the broadcaster in question are confined to one multiplex [10].

8.6.3   A look at past regulatory approaches for cable and satellite TV

To understand the starting position for regulation in the digital era, it is useful to look back at the policy objectives, developments, and reactions of past telecommunications and broadcasting policies [11].

8.6.3.1 Industrial policy for cable TV

With the intention to create competition in the traditionally highly regulated and mostly public service telecommunications and broadcasting sectors, the U.K. government in the early 1980s started a process of liberalization. This included the encouragement of the development of cable TV provided through private companies.

The initial development of cable was very slow. By October 1990, only 3.3% of U.K. households were connected to a cable-TV network, and growth rates were rather modest. This was due mainly to the fact that programming quality on cable (except for the four terrestrial channels) was seen as low. Cable operators faced high infrastructure costs so that little money could be invested in attractive programming. Hence cable did not offer an advantage over terrestrial TV, especially since cable became increasingly expensive for consumers.

In response to this poor development, in 1991, policies changed more in favor of cable. Operators were now allowed to offer telephony services in connection with cable TV, while BT and Mercury were still not permitted to transmit broadcasting signals over their telephone networks for at least another 10 years. While cable-TV penetration remained modest, revenues from the telephony services became more and more important for operators, and their market share increased rapidly.

As can be seen in Figure 8.1, after this policy change, the number of cable households has grown more rapidly. Since 1997, the number of new cable connections is even higher than that for new satellite

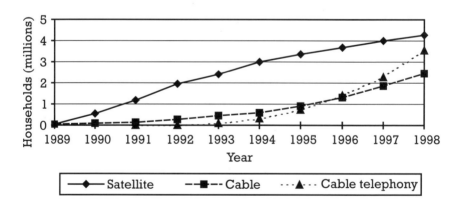

Figure 8.1 Cable- and satellite-TV households in the United Kingdom, 1989–1998.

connections. The possibility of dual revenue streams from one network also encouraged cable companies to extend their networks and pass more households.

Another policy change in favor of the cable industry was the admittance of investors from non-EU countries. As a result, in 1994, almost 90% of all U.K. cable-TV investment stemmed from North America, almost 70% from the United States. The main investors were four U.S. "baby Bells" as well as four large U.S. cable operators (TCI, Comcast, Jones, and Cox).

The introduction of satellite TV in 1989 surprisingly did not have a negative effect on cable. On the contrary, the heavy marketing campaigns of BSB and Sky raised consumer attention and interest in multichannel TV and increased consumer preparedness to pay for more channels. This also helped cable, in particular with the advantage to transmit the satellite channels over cable as well as offer attractive combinations of TV and telephone service.

8.6.3.2 Satellite policy in the United Kingdom

An important factor in the technology policy for the development of satellite TV in the United Kingdom was the 1981 Home Office study "Direct Broadcasting by Satellite." This study claimed that a national satellite industry was needed to secure a share in an internationally developing market, especially for direct broadcast satellites and information technology.

The government insisted that the necessary investments be carried out by the private sector. However, a consortium of the BBC, ITV, and private investors, named the Club of 21, after an initial planning phase decided that it was unable to finance such a project and, in 1985, discontinued.

This failure of an early development of a satellite industry in the United Kingdom had three major consequences. First, the BBC withdrew completely from any engagement in satellite (this was upheld until 1997). Second, the recommended early start of satellite (1986 was envisaged in the government study) could not be achieved. A start well before 1990 was seen as crucial to be able to compete with emerging foreign satellite offers. Finally, a central element of the industrial policy for satellite was lost: the objective of this industry to be national.

In December 1986, a license for three satellite channels was awarded to the BSB Consortium, which comprised Pearson, the ITV companies Granada and Anglia, the electronics company Amstrad, as well as Virgin.

The government protected the venture by delaying the license award for two other potential DBS channels until after the BSB launch. Moreover, the start of a fifth terrestrial channel was delayed until 1995. Despite these protective measures, however, BSB was not protected from competition. The government had disregarded technological progress. It had not realized, and therefore not been able to close that loophole, that low- and medium-powered satellites, which did not require a license, could be used for TV broadcasting.

With the employment of just such a satellite, Rupert Murdoch was able to circumvent the protectionist satellite policy in the United Kingdom. In February 1989, SkyTV, an unencrypted, advertising-financed four-channel satellite offering was launched, broadcasting from the low-power Astra satellite. With the employment of a matured technology, Sky gained a first-mover advantage, starting more than one year earlier than BSB, which employed a more sophisticated technology, but which therefore required more time to develop reception and transmission technologies.

In May 1990, when BSB launched, Sky started to encrypt its programs. In addition, an aggressive marketing campaign and subsidization of satellite dishes helped Sky to quickly establish a subscriber base of around 640,000 households.

Between May and October 1990, Sky and BSB competed fiercely, both losing large amounts of money in the bidding for content. In November 1990, the two competitors merged to become BSkyB.

After the merger, BSkyB established a dominant position in multi-channel TV. Cable did not provide serious competition to the satellite broadcaster. BSkyB had control over premium content and could thus oblige cable operators to broadcast its channels, thereby preventing them from setting up competitive offerings.

In summary, it can be stated that past regulatory measures for cable and satellite TV failed to achieve their intended effects in the United Kingdom: Contrary to government objectives, these industries are owned by non-U.K. (and non-European) companies and the market for pay TV is not characterized by healthy competition but is monopolized by BSkyB.

This failure of past industrial policy for cable and satellite TV has led to the belief that developments in pay-TV markets can only in a limited way be influenced by a strict application of competition regulation. Additional sector-specific regulation is regarded as essential. These experiences explain why U.K. regulators insisted on a detailed regulatory framework for digital pay TV to be established prior to the launch of services.

8.6.4 The regulation of digital TV

In the United Kingdom, due to its island characteristics, DTT appears attractive because border interferences are small. The U.K. government was the first in the world to have made available terrestrial frequency for digital-TV services and thus encourage existing and new broadcasters to make use of the six available blocks of frequency (multiplexes).

For the government, the incentive for supporting DTT is that available spectrum is used more efficiently due to signal compression. This means that higher revenues can be gained from the licensing of spectrum space to broadcasters or telecommunications operators, especially if spectrum is to be auctioned.

To foster the development of digital TV, the government declared its determination to ensure that all transmission media enter digital with equal chances and shares. The terrestrial platform was considered a disadvantage due to its (initial) limitation to around 30 channels and was therefore to receive special support.

8.6.4.1 The 1995 proposal by the Department of National Heritage

The first proposals for the regulation of DTT were published in August 1995 by the Department of National Heritage (DNH) [12]. This discussion paper is characterized by a fundamentally new approach to licensing. As seen, digital terrestrial frequencies were divided into six blocks, so-called multiplexes, each of which can transmit a minimum of three channels. The ITC was responsible for issuing the licenses for multiplex operators. The latter would then subcontract service providers to fill the multiplexes with content.

This provision had important consequences. By separating multiplex operators and program providers, the former would obtain a "gatekeeper" position between viewers and service providers. In addition to the provision of services, they could also control subscriber management, marketing activities, and the trading of frequencies between broadcasters. Taking into account the potentially powerful position of multiplex operators, the government suggested that no operator was to control more than two multiplexes. Applicants for licenses were evaluated according to their capability to invest in infrastructure and to finance advertising for the purchasing of digital receiver equipment and according to the spectrum of services they could offer.

The government proposal emphasized the importance of public service broadcasting and its participation in the development of digital transmission technologies. The incumbent national broadcasters BBC, ITV, Channel 4, and the new Channel 5 were guaranteed access to one-third of the available capacity, provided that at least 80% of their current programming would simultaneously be provided in a digital packet. However, the BBC was not allowed to finance new service provisions through revenues from license fees.

With regard to CA, the government acknowledged the fact that, due to network externalities and economies of scale, there should be few competing systems. BSkyB, with its existing (analog) pay-TV satellite system, was regarded as having an advantage in the development of such systems. To ensure nondiscriminatory access for program providers as well as multiplex operators, it is suggested that suppliers of access systems for digital transmission should be awarded licenses on the basis of the conditions laid down in the 1984 Telecommunications Act. It was to be OFTEL's task to monitor compliance with the licensing arrangements.

This proposal was received with much criticism. The 1995 proposal attempted to separate responsibilities for transmission and programs. This proved overly complex and impracticable. Moreover, the terrestrial broadcasters expressed strong reservations. The main point was that, although they would have access to digital transmission capacity, most advantages of digital broadcasting would be gained by the multiplex operators that can control marketing, decoding, and subscriber management. In addition, terrestrial broadcasters expressed their doubts about the profitability of digital services, implying that nobody would be prepared to substantially invest if there were no setting of a date for the termination of analog broadcasting. Furthermore, it was criticized that the readiness to subsidize equipment suppliers and manufacturers as a criterion in the licensing decision was given priority over the quality and variety of programming [13].

In response to the criticisms, the government permitted broadcasters to take control over multiplex operations. Furthermore, the appeal of program services to a variety of consumer tastes and interests was added as a criterion for license award. Still no date was given for the termination of analog terrestrial services.

The 1995 proposal can be regarded as the trigger for the public debate around regulatory issues of digital TV and, in particular, CA and EPGs.

8.6.4.2 The regulation of CA and EPGs

CA and EPGs are regarded as bottleneck facilities that can have an important impact on the relative success of competing broadcasters and, hence, would need to be regulated.

The debate around CA regulation   The key to CA regulation is the condition that all broadcasters should be able to gain "fair and reasonable access on nondiscriminatory terms" to CA technology, as laid down in the 1995 EU Directive. Of particular importance to operators is the clear definition of "fair and reasonable" and to what extent the subsidization of set-top boxes is considered to be anticompetitive.

Existing broadcasters expressed their fears that BSkyB, which announced plans to launch its digital service before the planned launch for DTT, would be able to assume a gatekeeper position by being first into the market with its set-top boxes. If these were incompatible with those for DTT and cable, then satellite TV would be able to dominate the market. The anticipated scenario provided an exact "rerun" of the situation at the launch of analog satellite TV in the United Kingdom, which subsequently led to BSkyB's monopoly in satellite TV (see Section 8.6.3.2).

The discussion was reminiscent of the proceedings of the DVB Group (see Section 5.3.3), but on a national basis. As the 1995 EU Directive had not resolved the issues of CA and set-top boxes, the existing broadcasters demanded appropriate legislation, so as to prevent a BSkyB digital monopoly on a national level.

A long and intense debate took place at both government and industry levels concerning the meaning of fair, reasonable, and nondiscriminatory access. The question was whether this implied the mandatory licensing of CA technology. The government finally decided that CA services, and not necessarily the technology, have to be offered to all interested broadcasters. If competition proves distortionary, the government can step in and mandate technology licensing. This is to be used as a last resort but at the same time serves as a warning to potential abusers.

Licensing policy for CA   Government legislation on CA came into force on January 7, 1997 [14]. CA was to be regulated by way of licenses issued by the DTI under the Telecommunications Act of 1984 and regulated by OFTEL. Companies wishing to run CA systems have to apply for a class license permitting them to provide CA technology. Furthermore, the licenses are to separate out encryption and scrambling services from subscriber management services, although the same company can hold licenses for both. The document further determines that any CA operator must provide sufficient information in advance to all interested broadcasters that will enable them to use the system. The CA operator must also agree to cooperate with any broadcaster in reaching an agreement to use the system.

OFTEL (1996) stated that under a class license a single operator would not be able to do the following:

- Use control of the CA technology to give it an unfair advantage against other broadcasters in competing for viewers;

- Ensure that its CA technology becomes the de facto standard;

- Be able to dictate terms to everyone else;

- License its technology in such a way that the set-top boxes would only work with one delivery system.

The document further stated that broadcasters would not be forced to package their channels with those of the CA operator and should be able to buy only the services they required from a menu of customer management services, subscriber management, subscriber authorization, and encryption.

The debate around EPG regulation   EPGs also form a bottleneck facility as their design can have an important influence on the success of a particular broadcaster. With regard to EPGs, OFTEL had already argued that market advantage could be achieved by a content or service provider controlling the navigation system and that the potential for distortion of the market, and therefore the need for regulation, would be even greater if a dominant network owner also provided content or services [15].

The 1996 Consultative Document suggests ways to prevent the EPG from being used to restrict, distort, or prevent competition between broadcasters. These included ensuring the independence and impartiality of a comprehensive EPG, separate EPGs for competing broadcasters' program packages, or ensuring equal access to the underlying information on program schedules. It was also proposed that the sequence of channel listings on the on-screen program guide should reflect the existing patterns of viewing, thereby preventing BSkyB from putting BBC and ITV channels at the lower end of the listing.

Detailed regulation for CA and EPGs   In March 1997, OFTEL published detailed regulations on CA. The provisions include details on pricing of CA services and set-top box subsidies and how a broadcaster using another broadcaster's CA services can retain commercial confidentiality of its subscriber base. It is further determined how EPGs can be made competitively neutral. Finally, regulation is provided on procedures enabling the operation of more than one smart card by competing broadcasters [16]. It has to be noted that these CA regulations apply to digital transmission only. Analog pay TV remains unregulated.

Parallel to OFTEL, the ITC published a code of conduct for the operation of EPGs. The code was drawn up in cooperation with OFTEL and incorporated the results of public consultations. The code emphasizes that viewers should have easy access to free-to-air channels and restricts levels of EPG branding by broadcasters. It also closes loopholes that could have exempted satellite text-based EPGs from regulation and extends protection to interactive services.

The issue of EPGs illustrates the existence of overlaps in the regulatory responsibilities of OFTEL and the ITC. EPGs are broadcast material—therefore an ITC responsibility—but also form a data (telecommunications) service that falls under the responsibility of OFTEL.

As argued earlier, those overlaps of responsibilities create inefficiencies and uncertainty for companies. A single regulatory body is likely to be significantly more efficient.

8.6.4.3   New regulation on channel bundling

With the increasing importance of pay TV and in response to complaints from cable operators and consumer groups, the ITC has adopted new regulations on channel bundling and minimum carriage requirements (MCRs). They came into effect in January 2000.

MCRs, demanded by program providers, serve to oblige program distributors (e.g., cable companies) to pay for and supply services to a minimum percentage of their customers. This is usually 80% to 100%. Contracts that include such MCRs result in viewers having to buy large packages of channels (i.e., channels are bundled). Moreover, a common practice in pay TV is to force viewers to buy a basic package (often consisting of around 40 channels) before they can subscribe to premium channels. Hence, viewers have to "buy-through" to premium content.

If viewers subscribe to several premium channels, they often receive free "bonus" channels. One of those was, for example, the Disney Channel, which until March 1998 was available only to customers who bought two BSkyB movie channels. It was a complaint from the cable industry on this issue that led the ITC to look into the issue of bundling.

The ITC argued that bundling slowed the take-up of satellite and cable services. The example of NTL seems to support this: The cable operator has been offering mini-packages since 1997 and, by the end of 1998, had achieved a penetration rate of 37% in its franchise area—around 10% higher than its nearest rival.

In June 1998, the ITC announced the abolition of MCRs from January 2000. Exempt from this are channels exclusive to one distribution medium, where those contracts were entered into before April 1998, as well as start-up channels that may negotiate MCRs for the first 12 months of their operation. Furthermore, all premium channels must be made available on an individual basis.

Cable operators have welcomed the ITC decision. For them, the abolition of MCRs implies the creation of a level playing field in the supply of pay-TV services. The ITC decision also appears beneficial for consumers as they will have more flexibility in choosing what they want to subscribe to. Independent program providers such as Flextech dispute this. They claim that viewer choice will be reduced because channels with a small audience share will be driven out of business. Without distribution guarantees (as part of a big basic package) they will not be able to generate sufficient revenues through subscription or advertising and sponsorship. From an economic point of view, however, this argument does not hold. The counterargument is that a viewer should not be forced to cross-subsidize another viewer's interests. And similarly, program makers should be able to reap the rewards from creating attractive content.

In reaction to the decision on channel bundling, BSkyB and ONdigital lodged complaints on the bundling of TV and telephony services. This move can be regarded as a "counterattack" on the cable operators. The main arguments are, first, that if channel bundling is anticompetitive, then TV/telephone bundling is as well. Second, they claim that phone line revenues can be used to subsidize TV operations.

An ITC/OFTEL consultation document of April 1999 states that the regulators have no objection in principle to the bundling of TV and telephony. Their view is that, as minor players in their respective fields compared with BT and BSkyB, the cable operators need to be able to use the economies of scale that their networks provide. Furthermore, consumers also get value out of such services.

8.7 Summary

The structure of the U.K. analog-TV market is quite similar to that in France, with the majority of viewers still watching a very limited terrestrial offering, an underdeveloped cable sector, some pay-TV penetration, and one dominant pay-TV operator. This suggests that there is potential demand for digital multichannel services.

Existing broadcasters, however, were reluctant to launch digital TV. This was due to the high investments needed and uncertainties with regard to market developments and the regulatory framework. BSkyB held back initially due to uncertain market conditions and in order not to jeopardize its analog pay-TV operations. It was the government that provided incentives persuading existing broadcasters to launch DTT. This created the stimulus also for BSkyB and the cable companies to launch digital services.

8.7.1 Competition has created rapid penetration

The United Kingdom is very specific in that four digital pay-TV services have been launched on three different distribution media: satellite, terrestrial, and cable. Nearly simultaneous launches have resulted in head-to-head competition without any significant first-mover advantages. All operators are aiming to quickly establish a large base of subscribers, and, to this end, heavy investments have been made. The rapid take-up of digital services is illustrated in Figure 8.2.

The importance of equipment subsidies is illustrated by the even higher take-up rates once set-top boxes were given away free of charge.

It is certain that the fierce competition resulted in rapid penetration. The services offered are attractive to consumers: Switching costs are minimal, and channel packages have become more flexible than in analog pay TV, while the number of channels has been greatly increased and prices have come down.

Due to the different characteristics of the distribution media and differing consumer preferences for those and for channel line-ups, it appears possible that all three platforms will be able to maintain a profitable market share in the long run. It is likely though that ONdigital will have a smaller share than cable and satellite services. The competition between SkyDigital and digital cable services has not really started yet. Industry experts expect that, after about five years, cable will become the leading supplier of digital TV. The media agency Optimedia, for example,

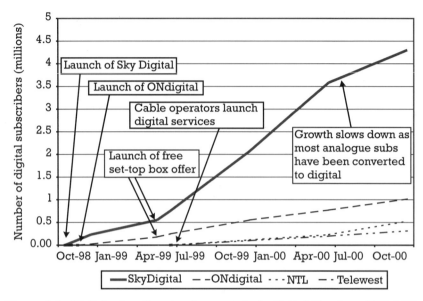

Figure 8.2   Digital subscriber developments in the United Kingdom, 1998–2000.

estimates that by 2003, 53% of U.K. homes will receive digital TV, with SkyDigital and digital cable accounting for around 20% each. ONdigital will have 7%. From 2004, they expect cable to take the lead.

As has been examined in Section 2.3, lock-in strategies are pursued in the hope to recoup investments (such as subsidized equipment) in a later period. Providing free set-top boxes may not achieve lock-in since after one year (when the minimum subscription period expires), consumers can easily switch to a different provider. This trend is increasingly observable, for example, in the mobile phone industry. To avoid churn after one year, the operators will have to compete on content and customer service or offer loyalty schemes. That is why all operators stress their program extensions (new channels, new interactive services), pricing offers, and program customization (smaller packages). Much will also depend on the business strategies in this second period: Will boxes still be free? Will idTVs be subsidized? How will package prices evolve? Which loyalty schemes will be offered? Loyalty schemes could involve discounts for second-year subscriptions, increased program flexibility for long-term customers, or even new offers where no 12-month contract is needed.

8.7.2 Regulation created certainty for digital-TV players

A reason for the initial reluctance of broadcasters to move to digital is the regulatory uncertainty that prevailed while the government and different regulatory bodies discussed the detailed interpretation and implementation of the provisions given in the 1995 EU Directive.

In contrast to Germany, France, and most other European countries, the debate on the regulation of digital TV in the United Kingdom has been intense. This is strongly a reflection of past regulatory approaches and the resulting market structures, particularly the dominant position of BSkyB. The introduction of digital TV was regarded as a chance to re-establish a level playing field for broadcasters on all distribution media. BSkyB's dominance in satellite TV and its control over the cable operators' access to programming put the issue of this company's potential dominance in digital TV on the political agenda. The BBC and the ITV companies, which saw a chance to change the conditions with the arrival of digital TV, attempted to convince the regulators that similar market structures have to be avoided at all costs in the digital era, especially in light of the fact that BSkyB's dominance in analog pay TV would automatically give it a head start in digital.

Prices charged by BSkyB to consumers and intermediaries (cable and DTT operators or content providers wanting to use the BSkyB distribution platform) are supposed to be fair and nondiscriminatory. The responsible regulatory authorities have put in place detailed regulations on "bottleneck facilities" (i.e., CA and EPGs). It remains to be seen, however, whether this regulation is appropriate to resolving disputes. The determination of "fair and nondiscriminatory" prices is rendered almost impossible due to the lack of information on true costs of service provision by digital operators.

In addition to the regulation of bottleneck facilities, the U.K. regulators have also abolished market distortions that existed in analog pay TV. Those are, in particular, channel bundling and minimum carriage requirements. This improves the situation of cable operators, which can thus engage in effective competition with BSkyB.

In summary, it can be stated that U.K. regulators have adopted several measures to encourage and support the development of competitors to BSkyB in the digital era while at the same time safeguarding consumer interests (i.e., the programming advantages of BSkyB). So far, it appears that competition is healthy. In particular, the conditions for cable operators have improved significantly. In fact, they are even expected to eventually assume market leadership.

Strong points of critique on the U.K. regulatory system are the unclear structures and responsibilities of regulatory bodies. This has led to delays, duplication of work, and competition among regulators. This creates uncertainty for companies and is likely to have delayed the introduction of digital-TV services. The development of digital TV as well as other new multimedia services and convergence trends in general reveals the need for a fundamental reform of the regulatory system for media in the United Kingdom.

Further assessment, notably in comparison with Germany and France as well as with regard to the earlier theoretical discussion, can be found in Chapter 9.

Endnotes

[1] Hill, J., and M. Michalis: "Digital TV and Regulatory Issues: The British Case," *Communications & Strategies*, No. 27, 3rd quarter 1997, pp. 80–83.

[2] Hill, J., and M. Michalis: "Digital TV and Regulatory Issues: The British Case," *Communications & Strategies*, No. 27, 3rd quarter 1997, p. 85f.

[3] Hill, J., and M. Michalis: "Technological Convergence: Regulatory Competition. The British Case of Digital TV," *Policy Studies*, Vol. 18, No.3/4, pp. 219–237.

[4] Quoted in Kavanagh, M., "Battle Lines Are Drawn in U.K. Digital TV War," *Ad Age Creativity*, October 5, 1998, p. 1.

[5] Commission of the European Communities, "Directive 95/47/EC of 24 October 1995 on the Use of Standards for the Transmission of TV Signals," *Official Journal of the European Communities*, November 23, 1995.

[6] Commission of the European Communities, "Directive 95/47/EC of 24 October 1995 on the Use of Standards for the Transmission of TV Signals," *Official Journal of the European Communities*, November 23, 1995.

[7] Cave, M., and M. Shurmer, "Business Strategy and Regulation of Multimedia in the United Kingdom," *Communications & Strategies*, No. 19, 3rd quarter 1995.

[8] Department of Culture, Media and Sports, *Green Paper on Regulative Communications*, CM 4022, London, 1998.

[9] Department of National Heritage, *Guide to Media Ownership Regulation*, London: HMSO, 1996, paragraph 7 of Part III of Schedule 2.

[10] For detailed analyses on these, see, for example, Goodwin, P., "Television Under the Tories: Broadcasting Policy 1979–1997," British Film Institute,

London, 1998; and Davis, J., et al., "Industrie-politische Wunschvorstell-ungen bei Kabel und Satellit Gescheitert," *Media Perspektiven*, No. 6, 1998, pp. 298–309.

[11]   Department of National Heritage, *Digital Terrestrial Television. The Govern-ment's Proposals*, London: HMSO, 1995. Note that in 1997, the DNH was renamed Department of Culture, Media and Sport (DCMS).

[12]   ITV, *The ITV Response to the Government's Proposal for the Introduction of Digital Terrestrial Television*, October 1995.

[13]   Department of Trade and Industry, *Class licence for the running of telecom-munications systems for the provision of conditional access services granted by the Secretary of State under Section 7 of the Telecommunications Act 1984 on 7 January 1997*, London, 1997.

[14]   Office for Telecommunications, *Beyond the Telephone, the Television and the PC*, London, August 1995.

[15]   Office for Telecommunications, *The Regulation of Conditional Access for Digital Television Services, OFTEL Guidelines*, London, March 1997.

[16]   Quoted in *Marketing*, "Digital TV development in the U.K.," July 22, 1999, p. 7.

CHAPTER

9

Contents

Comparative analysis of market developments in France, Germany, and the United Kingdom

9.1 Introduction

This chapter analyzes the success factors for rapid digital-TV penetration by comparing the market developments in France, Germany, and the United Kingdom.

The existence of common standards and the way they had been achieved formed the basis for several players quickly launching digital-TV services, thus creating a competitive environment early on.

Despite the overall success of the DVB technologies and their availability on the European markets, the actual implementation of digital-TV services has shown very different results in the three countries studied. Success has been particularly influenced by three factors: first, the existing market structures and pay-TV services; second, strategies of digital pay-TV operators and their propositions to consumers; and third, the regulatory framework and the public debate around the regulation of digital pay-TV services. These factors will be examined in this chapter with reference to the theoretical discussions of Chapter 2.

9.2 The impact of standardization on digital-TV penetration

An important advantage of standardization is that it creates a level of certainty and security for market players. Rather than developing and investing in a new technology alone, without knowing whether this technology will survive in the longer term, standardization reassures players that a technology will stay. The fact that several players introduced services shortly after the standardization process was finalized illustrates this. Furthermore, in a standardization effort, many players share the costs for developing standards. In this way, cost for each player is reduced, risk is reduced, and, in a concerted effort with input from many sides, it is more likely that the best technology is chosen. In the case of DVB, widespread international acceptance of the DVB standards suggests that the technology is sound.

A further positive effect of standardization is that it is beneficial to effective competition. Once the DVB standardization process was in progress, no player announced to launch digital TV on its own, trying to preempt the market. All waited for the outcome of standardization to then launch services. This creates a level playing field in which several players launched services within a relatively short time span. This helped to create effective competition as no one could gain a significant first-mover advantage. The competition established in the United Kingdom and in France illustrates this.

The standardization process also helps to create awareness of new technology developments among industry players as well as government authorities and, possibly, consumers. The DVB group invited all interested parties into the process, thereby also ensuring that it was not viewed as a cartel. Government authorities were made aware of the new technology developments. In the case of the United Kingdom, this had the effect that policy makers were prepared for the launch of digital services. They created incentives for the launch of new services and provided a regulatory framework, thus supporting a competitive environment.

The standardization of transmission standards for the whole of Europe and for all transmission media (cable, satellite, and terrestrial) is likely to have had a positive impact on cross-border cooperation as well as competition. In the past few years, we have seen a number of cross-border partnerships, such as BSkyB taking a stake in Premiere World, France Telecom investing in NTL, or Canal Plus expanding into Scandinavia, Italy, Belgium, and the Netherlands. In analog TV, a number of different standards prevail in different European countries (Secam, PAL, and subtypes of PAL), inhibiting effective cross-border competition. Hence, the creation of pan-European standards has helped to improve competition in the region as a whole.

The fast standardization and implementation of digital technology in Europe has also had an important international economic effect. The open standards, shown to be working in several European countries, proved popular among non-European countries. DVB standards are already being employed in more than 60 non-EU countries. This creates new market potential, particularly for European manufacturers, but it also has an important political effect: It suggests that Europe has developed a de facto world standard in an important area of information and communications technologies.

The fact that CA was not standardized was, in 1995, by many seen as a failure of the standardization process because it would lead to consumers being locked into a proprietary technology. However, it was the only way to provide an incentive for existing pay-TV broadcasters to develop digital-TV services. Players such as BSkyB and Canal Plus were the ones prepared to invest heavily to develop digital-TV services. To do so, however, they had to be reassured that they would be able to reap the rewards from their investments through deployment of a proprietary CA technology. Today, it can be seen that lock-in of consumers has not taken place. At the same time, the initiatives of existing pay-TV broadcasters have stimulated the market for digital-TV services, in particular in France and the United Kingdom, but to a lesser extent also in Germany. Healthy competition has evolved in France and in the United Kingdom. It is based mainly on program and service offerings and has created favorable conditions for consumers. Lower prices and better service have contributed to fast penetration of digital TV. Germany forms an exception, but this is due to its particular market structures and player strategies.

In summary, it can be noted that standardization has created favorable conditions for fast market penetration. However, it does not guarantee success—as can be seen in the German case. A number of other factors influence the take-up of digital-TV services. Section 9.3 discusses these.

9.3   Preconditions for the launch of digital TV

Existing market structures determine the added value for consumers of new digital pay-TV services. Therefore, they have an impact on the take-up of those.

9.3.1   Penetration levels of multichannel and pay TV in France and the United Kingdom

France and the United Kingdom display many similarities in analog-TV markets. In both countries, the large majority of households (87% and 75%, respectively) receive a very limited terrestrial offering of around five channels. Both countries have a low cable penetration of around 10% of households. In addition, cable operators are largely dependent on programming content from the main broadcasters and, in particular, the dominant pay-TV operators. The satellite sectors in both countries are fully under the control of the dominant pay-TV operators. In the United Kingdom, BSkyB runs analog satellite pay TV with a subscriber base of 3.6 million (16% penetration) whereas in France satellite TV is received only by around 6% of TV households.

Valuable insights can be gained by looking at the penetration of pay-TV services. In France, pay TV has been established for more than 10 years with Canal Plus reaching most of its customers via terrestrial and, to a lesser extent, cable and satellite transmission. In the United Kingdom, pay TV provided by BSkyB is received mainly by satellite but also via cable. Terrestrial pay TV is not offered. This suggests that there is potential for terrestrial pay TV in the United Kingdom.

With a penetration rate of 27% in France and 23% in the United Kingdom (1997 figures), pay TV in both countries is well established. This implies that consumers are generally prepared to pay more (in addition to license fees) for multichannel and premium programming. Since digital-TV services, as explained in Section 4.4.1, will mainly be provided by means of pay TV, it forms a significant advantage if consumers are already used to paying for program packages, single channels, or even single events. The threshold for paying is likely to be lower.

The fact that most households still only receive a small terrestrial channel offering suggests that multichannel TV has a significant attractiveness in these two countries. This is especially the case in France where even pay TV has traditionally not offered multiple channels, but for a long time consisted of only one additional channel, Canal+, at the relatively high price of F220 per month. In the United Kingdom a basic package of around 30 channels could be had at just over half of that sum (see Table 6.2). Hence, in France a lower priced offer has large potential. This is the strategy that is successfully pursued by TPS.

9.3.2 Reputation and brand of existing broadcasters in France and the United Kingdom

An important factor for the successful introduction of new services is the existing reputation and brand of broadcasters. The dominant pay-TV providers BSkyB in the United Kingdom and Canal Plus in France achieved their success in analog pay TV through a reputation for quality content, particularly the provision of premium sports and movies for which they often hold the exclusive rights. This amounts to a significant competitive advantage for the introduction of digital pay-TV services.

BSkyB has been known for years in the U.K. market as a large and successful company under the leadership of Rupert Murdoch, an important international player in audiovisual and print media. With this notion in mind, consumers are likely to trust that BSkyB will be able to enter the digital pay-TV market equally successfully. Moreover, through its powerful position in U.K. media and its resulting influence on public opinion, BSkyB can make sure that it is presented in a favorable light and at the same time can create an awareness and a positive attitude toward its new services. Thus, there is a real advantage for the powerful media tycoon in being an incumbent in media and pay-TV markets. This argument will be picked up again in Section 9.4.

In France, it is the Canal Plus brand name rather than that of a particular person that is widely known. However, here too, the well-known brand associated with a powerful and successful TV company with access to large film libraries and involvement in film production can create confidence among consumers that if it decides to enter the digital market, it is likely to become a success. In this sense, the existence of strong brands can lead to the effect of a self-fulfilling prophecy: digital pay TV becomes a success because consumers are convinced it will become a success and therefore adopt the new technology.

The second French digital pay-TV operator, TPS, can profit from the brand names of its members which consist, among others, of the nationwide free-TV broadcasters TF1, France Télévision, and M6. The involvement of public broadcaster France Télévision in particular can be regarded as an important asset, as public services are regarded as a guarantee for quality content and thus enjoy consumer confidence and popularity. The second digital operator in the United Kingdom, ONdigital, can build mainly on the reputation of the nationwide analog commercial broadcasters Carlton and Granada, which are also popular, although they do not have the same status as public broadcasters.

In summary, it can be said that the market structures in France and in the United Kingdom were in general favorable for the introduction of digital multichannel pay-TV services. The situation in Germany, however, is very different.

9.3.3 Preconditions are very different in Germany

Although the German market for television is very large and potentially lucrative, the launch of digital pay-TV services has been nothing less than a failure. This failure has a number of explanations. To begin with, the German market structure is considered, and its conditions are shown to be, very different to those prevailing in France or the United Kingdom.

A hindrance to the attractiveness of pay TV in general, frequently quoted as the main reason also for the failure of digital pay TV in Germany, is the large free-TV offer combined with the very high penetration of cable and satellite. Although, as in France and in the United Kingdom, the analog terrestrial offering is very limited, only around 30% of households opt for this distribution medium. The majority of households are connected to either satellite (20%) or cable (50%) receiving a free (mostly advertising financed) multichannel offering consisting of around 30 channels. (There is, however, a monthly fee of around DM26 to be paid for the cable connection.) This implies that paying extra for multichannel TV has no great appeal to German viewers. The experience of ever-more channels with increasingly poor content quality and frequent interruptions for advertisements has even led to the public broadcasters regaining viewer share. On the other hand, this situation suggests that there might be demand for high-quality (premium) content with no advertising. The experience of the Premiere channel, however, raises doubts.

Existing analog premium pay TV has not proved popular with German consumers. The Premiere channel attracted only around 1.3 million customers over five years. This implies a penetration rate as low as 3.5% of all TV households and is in sharp contrast to the penetration rates in France and the United Kingdom. Consumers seem to perceive the price of DM50 as expensive for just one additional channel in the large range of freely available channels. Not even the fact that certain sports and movie events are exclusive to Premiere can persuade many consumers to subscribe to the pay-TV offering. The digital offer Premiere World appears to be suffering the same fate.

Another disadvantage to the successful launch of digital TV in Germany is the fact that there exists no distinct brand name associated by consumers with a company that provides quality premium services. Premiere, which could have aimed to develop such a brand, was (until recently) owned jointly by three companies (CLT-Ufa, Canal Plus, and Kirch) that are competitors in other areas of business. This is likely to have prevented any single one of them being promoted as the brand name behind Premiere.

In general, German TV channels are run under channel-specific brands, but not under the name of the controlling company. That is why few consumers are aware of the fact that commercial television is controlled by two large groups, Bertelsmann and Kirch. With this lack of an existing brand and reputation for the introduction of a new service it is much harder to create consumer awareness and confidence in the quality and durability of a new service. The Bertelsmann name, in particular, is well known for its activities in audiovisual as well as print and on-line media in Germany. Similar to BSkyB in the United Kingdom, it could have provided a strong brand for the introduction of digital pay TV.

In response to the existing market structures, an entry strategy has to be devised very carefully so as to provide a significant increase in value for viewers and incentives to switch to digital. Section 9.4 will analyze the strategic approaches of operators.

9.4 Strategies and consumer propositions for digital services

This section compares the strategies and consumer propositions of the digital pay-TV operators in the countries analyzed.

Most importantly, new digital services have to offer a significant added value over existing services so as to overcome switching costs. Furthermore, as argued in Section 2.3, the introduction of a new service with a potential "lock-in" for consumers (through the CA system) will have to offer some initial discount to consumers to compensate for this danger. Alternatively, an operator could deploy a strategy that prevents a lock-in for consumers. This would mean, for example, using a decoder with an open CA system or renting decoders rather than selling them.

9.4.1 The first movers
This section compares the strategies of the companies who were first in their respective markets to launch digital services: Canal Plus, BSkyB, and DF1.

9.4.1.1 Canal Plus: Added value without lock-in
The Canal Plus strategy corresponds well with the factors identified above as well as with the existing market structures. Most importantly, Canal Plus prevented fears of "lock-in" for consumers by making its decoder available for rent at a low monthly fee (F45) from the start. This reduces considerably the threshold for viewers to switch to the CSN service as they are aware that they could easily cease subscription or change provider without incurring significant costs. Consumers are bound temporarily through a 12-month minimum subscription period, but this is not regarded as a "lock-in" in terms of network economics.

The launch offer of CSN was also devised in a way that consumers perceive as added value. First of all, Canal Plus acquired rights from several U.S. majors and holds exclusive rights for popular sports events. These acquisitions, being made and publicized around the time of the digital launch, combined with the past reputation for premium content will have inspired consumer confidence. The intention of Canal Plus to expand the digital offering by adding new channels was also well publicized.

Second, the digital service offers a basic 14-channel package including the Canal+ channel at a price (F155) significantly lower than that of only the Canal+ analog channel (F220).[1] This offering is also larger and cheaper than the multichannel offering available on analog cable, which offered 10 thematic channels at F90 to F175. Especially aimed at persuading existing Canal+ subscribers to switch to digital, the premium channel (in combination with the basic package) is offered at a discount to the analog one and is shown on four channels at staggered starting times. Furthermore, a range of new premium channels to choose from as well as pay-per-view is offered. This large program offer responds well to the potential for multichannel TV in France and can be seen as one of the key drivers of the CSN service.

---

1. Although channel numbers offered and prices charged have changed since launch, the offers at launch or shortly after are used for reference here as those are the ones decisive for market entry and the acquisition of an initial installed base of subscribers.

9.4.1.2 BSkyB: Consumer choice but initial lock-in

BSkyB in the United Kingdom adopted a similar strategy with regard to the channel offering. However, as multiple channels were already available in analog pay TV, BSkyB needed to differentiate SkyDigital further and so decided to offer around 140 channels on its digital service. The marketing message suggests to viewers that there are special interest packages for a large variety of tastes. SkyDigital offers a selection of small basic packages that already provides more choice to the viewer than the analog service. Premium channels can be subscribed to one-by-one or in packages, then at reduced prices and/or with added bonus channels.

In general, the BSkyB proposition is characterized by a large offering with significant choice for viewers at prices on average lower than those for analog pay-TV. SkyDigital constitutes an immense proliferation of channels, but at the same time gives viewers more choices from which to select their preferred package. This provides an advantage over the analog pay-TV service, which had only one basic package and a limited number of premium packages on offer. The relatively low prices and the higher level of consumer control over channel choice are aimed at attracting new subscribers who might initially not want to spend much on pay TV but are interested in trying out a small number of special interest channels.

A fundamental difference to the Canal Plus strategy is the fact that BSkyB initially only sold decoders. Over the first seven months from launch, BSkyB aimed to lock in consumers. To mitigate consumer fears in this area, the decoders were subsidized. Moreover, consumer research has shown that the subsidized price of £200 is a "threshold": the highest possible price that a majority of consumers would be prepared to pay for a new consumer electronics device. This threshold price has been found to be similar also for most other European countries.

In addition to the decoder subsidization, BSkyB offered further incentives to potential customers. Free installation of the digital satellite equipment (costing around £80 per subscriber) and the possibility of paying in installments for the decoder are provided. This, again, helps to lower the entry threshold to pay TV for new customers. Existing BSkyB subscribers were offered additional incentives to switch to digital with a reduction in programming costs and a reduction in decoder price (£159 instead of £199 for new subscribers) and a reduced connection fee (£25 instead of £30) compared to new customers.

The strategy of lock-in but with compensation offers for consumers proved successful initially: 350,000 subscribers could be gained over four

months. However, with increasing competition from DTT and the pending launch of digital cable, subscriber growth slowed and the strategy was
altered. From June 1999, BSkyB started to give decoders out at no charge
in return for a one-year subscription. This complete subsidization of
equipment can prevent consumer fears of lock-in. Together with attractive content and pricing, the overall consumer proposition is even more
popular now, resulting in rapidly rising subscriber numbers.

The subsidies for the equipment are provided through Open, a consortium of well-known, reputable companies from different sectors. This
move was designed to create confidence in the future availability of the
service, especially as these companies were prepared to invest heavily in
the set-top boxes and planned to develop new interactive services. This
strategy of gaining consumer confidence seems to have worked in the
early stages. Once subscriber numbers were growing rapidly, it proved
more advantageous to BSkyB to take over the majority of Open.

The BSkyB strategy is characterized by fast reactions to competitive
pressures. Incentives such as free boxes, reduced telephone rates, free
Internet access, and the distribution of boxes in additional retail outlets, is
aimed at matching the competitive advantages of cable operators. Smaller
and more flexible program packages and the announcement of an idTV
for both DTT and digital satellite reception aim at matching the offers of
ONdigital.

BSkyB can follow such a strategy because it has the financial means
and strong allies that enable it to provide the high investments required.
It could be argued that if consumers are not locked in to BSkyB's system,
the company may not be able to recoup its initial investments. It is likely,
however, that BSkyB will be able to sustain its position in the long run
due to its competitive advantage in terms of content.

9.4.1.3 DF1: Little value added and high costs for consumers

In contrast to the strategies of the French and U.K. first movers, the
approach of DF1 in Germany clearly illustrates several shortcomings.
First of all, the DF1 offering failed to provide (and communicate) real
added value to viewers. The channel offering of an initial 14 thematic
channels combined with a very limited premium offer is not a significant
improvement over the existing free-to-air services available, despite the
price of the package being relatively low at around DM20 per month.

The most important hurdles to the successful introduction of DF1 are
not the monthly fees but the high initial costs. As in the United Kingdom,
DF1 set-top boxes, at least initially, had to be purchased by consumers.

Kirch subsidized the boxes, but at a retail price of DM890 they are considerably more expensive than the BSkyB boxes and are beyond the above mentioned consumer "threshold" price (around DM500). Moreover, existing satellite dishes had to be upgraded at an additional cost of around DM150, thereby increasing the total initial costs for consumers to over DM1,000. Considering this high upfront investment needed, combined with the perceived risk of lock-in and the limited program offer, it is not surprising that few people were attracted to DF1. Additionally, electronics retailers reported several technical problems with the d-box.

The Kirch name was not a strong brand in the market and the DF1 service was not connected to any well-known brand. Moreover, Kirch did not gain any important, reputable partners (as BSkyB did) for the further development of the service and the subsidization of the decoders. Lack of reputation and brand is less conducive to the creation of consumer confidence. Moreover, the rumors of the termination of DF1 in 1997 (after the merger between Premiere and DF1 had been blocked) created uncertainty among consumers as to the future availability of the service.

Another crucial shortcoming in the DF1 market entry strategy is Kirch's failure to gain carriage on the cable-TV network from the start. In this way, he did not have access to the large majority of German viewers who could have switched to DF1. However, it is much more difficult to persuade a cable customer to switch to satellite TV. Again, the lack of ensuring support from strong partners, in this case Deutsche Telekom, creates a disadvantage.

Similarly, Kirch did not have access to existing pay-TV subscribers (i.e., Premiere subscribers). As those already show a demand for pay TV, they are predestined customers and more easily converted to digital services. In France and in the United Kingdom, former analog pay-TV subscribers constitute a large share of early subscribers to digital pay TV; hence, they form an important market segment in the market entry phase.

Kirch attempted to create consumer confidence through heavy investments into content (mainly U.S. movies and sports) and by committing himself to acquire one million d-box decoders from Nokia. However, in the light of much content being available already in analog free TV, this did not convince viewers. The disappointing subscriber numbers two years after launch illustrate this.

It would appear that Kirch rushed into the market to gain a first-mover advantage but failed to deploy a differentiation strategy needed to gain an installed base. He attempted to win a mass market but did not ensure access to as large a customer base as possible (i.e., cable customers). He attempted to lock in consumers but did not offer them significant compensation for carrying that risk. Most importantly, he failed to provide a service with real added value.

As a result of the poor performance of DF1, strategies were subsequently altered. One year after launch, the d-box was made available for rent. The program offering has been extended to 32 channels (20 in the basic package). In November 1997, Kirch gained access to part of the cable network. Furthermore, consumers could "test" all DF1 channels for three months at DM50 before deciding which package to subscribe to. These modifications, however, did not significantly improve the situation. Eighteen months after launch, DF1 had only 280,000 subscribers.

It became clear that DF1 could not survive as a stand-alone service. That is why Kirch decided to take over Premiere and merge the two pay-TV operations. He also decided to make his decoder technology, formerly exclusively licensed to Nokia, available to other manufacturers. In this way, it becomes more likely that other broadcasters would also use it; hence, the CA system embedded in the d-box can become the de facto standard in Germany. Three years after the initial launch, Kirch relaunched digital pay TV in October 1999, now combining DF1 and Premiere to Premiere World. The program offering and pricing, however, are not much different.

It can be summarized that Kirch's aggressive first-mover strategy targeted at locking in customers did not offer sufficient compensation to convince consumers. Even when lock-in fears were reduced (through decoder rental), the digital pay-TV offering was not sufficiently attractive to present a real added value over analog services.

9.4.2 The second movers

The strategies of the second movers will be compared particularly in regard to how they differ from those of the first movers as this is a crucial success factor for them.[2] For second movers, again, existing market structures have to be considered, but additionally, they are forced to adapt their strategy to those of the first mover. In particular, they have to consider the lock-in choice of the first movers. For example, it is very

---

2. "Second movers" comprise all followers to the first mover.

unlikely that a second mover could succeed in selling decoders if the first mover is renting them. Hence, the second movers are limited in their strategic choice.

9.4.2.1 TPS: Differentiation on price and programming

The French case illustrates that systems competition can accelerate market penetration of a new technology—provided that adequate strategies are deployed. Furthermore, economists and industry experts predicted that only one operator could survive in each national market. In France, however, two platform operators serve the market and have sustained their positions for more than five years. Important here is that they offer sufficiently differentiated services.

The most successful case is the strategy pursued by TPS in response to Canal Plus. TPS decoders are rented, and, to offer added value over analog, a multichannel offering is provided with a focus on movies. This option is even available independent of a basic package and thus responds well to the lack of movies on free TV. TPS mainly differs from CSN in price. While the basic package is only slightly cheaper than that of CSN, the premium bouquets offer reductions of up to 48% compared to Canal Plus (see Table 6.5). This price differentiation also forms a significant improvement over analog pay TV, where the premium channel Canal+ has traditionally been relatively expensive. Overall, the strategy responds well to market demands in France. Furthermore, TPS might be able to draw in subscribers of the analog Canal+ channel who want premium content but are price-sensitive. A further differentiation from the Canal Plus digital offering is TPS' exclusivity of several terrestrial free-TV channels. In particular, the public channels France 2 and France 3—but also the popular TF1 and M6—are likely to have attracted viewers to the TPS bouquet. As mentioned earlier, TPS also benefits considerably from the reputation of these broadcasters that are the shareholders in TPS. With these widely known brands, supported by a F150 million advertising campaign, the new service proved very successful in raising consumer awareness and a reputation for quality.

Another strategy to differentiate TPS is its focus on interactive services, which offer something genuinely "new" for television. It has to be said though that Canal Plus, too, increasingly provides interactive services. Hence, it is more the advertising and promotion for these services that is crucial rather than their actual uniqueness. The fact that TPS interactive services induced Canal Plus to develop similar offerings shows the positive effects of competition. Under competitive pressure, with each player trying to offer more and more sophisticated services, this is fostering a faster development process, faster market penetration, and improved consumer acceptance of interactive services.

In summary, it can be stated that TPS has very successfully deployed a differentiation strategy. The company succeeded in establishing a stable market presence alongside the first mover, although the latter had been claimed to have so many competitive advantages that no competitor could have a chance. Among those are existing subscribers, expertise in pay TV, premium content, and control over technology. TPS proved many industry experts wrong and illustrates that there is space for a second operator in a national market.

The fact that both broadcasters, Canal Plus and TPS, chose not to attempt to lock in consumers through decoder sales is a significant factor for fast market penetration: First of all, the adoption threshold for consumers is considerably lower, and second, the competitors have made continuous efforts and improvement to their service offerings so as not to lose viewers to the competing service. For the market in general, this implies more rapid penetration with expanding channel offerings, responding to consumer demand, and at prices that reflect healthy competition.

9.4.2.2 ABSat: A too small program offering

In contrast to TPS, ABSat could not maintain its position as a platform operator. ABSat followed a niche strategy but could not gain a sufficient number of subscribers to sustain its stand-alone service. ABSat has instead become a program provider to Canal Plus and TPS. It might be argued that there is no space for more than two operators in the market. However, it might also be that ABSat's channel offering was too small and comparatively expensive to find a large enough market. ABSat can, however, exist as an independent program provider (similar to Flextech in the United Kingdom) supplying content to the digital platforms of the large broadcasters.

9.4.2.3 French cable: Complementing satellite offers

The cable operators in France do not really constitute competition to the digital satellite operators. First, they are part-owned by either Canal Plus (NC Numéricable) or TPS (Lyonnaise Cable, France Telecom Cable) and are largely dependent on content from these. Second, their digital offerings are relatively expensive and, as in analog, they could not gain large subscriber numbers. Finally, they do not yet make large use of cable-specific advantages, such as interactivity, telephony offers, and Internet access.

9.4.2.4 ONdigital: Differentiation through means of transmission

As noted earlier, in terms of existing market structures, the United Kingdom and France appear very similar. ONdigital launched only very shortly after SkyDigital, whereas TPS launched eight months later than CSN. This can be regarded as an advantage for ONdigital, as BSkyB could not establish a large first-mover advantage.

The basic elements of the strategies of the two U.K. competitors are similar. Both companies initially sold decoders at the price of £200 and then started to give them out at no charge. Both companies also offered additional incentives such as reduced-rate telephony, Internet/e-mail, and interactive services. Hence, both companies aimed at matching the other's offers—and the prospective offers from cable operators.

In terms of package prices, the competitors had similar offers too. Basic packages (of five or six channels) start at around £7, and one premium channel (including a basic package) can be purchased from around £18 per month. In contrast to TPS, ONdigital does not offer a price differentiation. Instead, the company aimed to differentiate itself by advertising its "pick 'n mix" approach for choosing customized packages. This does, however, not form a real differentiation, as SkyDigital, too, offers a range of basic bouquets and any combination of premium channels.

Most importantly, ONdigital has a significant disadvantage in terms of channels available. In contrast to 140 channels on SkyDigital, ONdigital has capacity only for around 36 channels. In this light, the ONdigital offers cannot compete: In comparison, subscribers get a better choice at a lower price with SkyDigital, which offers a range of predetermined packages, often enriched with bonus channels.

A further disadvantage for ONdigital is with regard to interactive services. The limitations of terrestrial transmission do not permit many such services and will never be able to offer real interactivity (i.e., a direct return channel independent of the telephone network).

Similar to TPS, ONdigital aims to differentiate itself through exclusive channels. ITV and ITV2 are available only on DTT. It can be expected that these channels are attractive to viewers, as ITV is an established channel with high popularity, and ITV2 will soon be screening Champions League soccer. It remains to be seen, therefore, if two exclusive channels will be decisive in making consumers subscribe to ONdigital. In France, exclusivity has proved popular. The TPS offering initially included four popular channels exclusively, two of which were public broadcasters, and this is regarded as an important contributor to developing TPS market share.

The premium channels offered on ONdigital are mostly provided by BSkyB and are all available on SkyDigital, too. Hence, ONdigital does not offer anything different in terms of premium channels, but instead serves as promotion for SkyDigital: ONdigital viewers getting a taste of BSkyB content and wanting more channels are likely to upgrade to SkyDigital, especially as they do not face any switching costs since decoders are free.

ONdigital can avoid such a trend only by offering more channels itself. Channel proliferation on digital terrestrial, however, is limited and depends essentially on the government's decision to terminate or fade out analog terrestrial TV so as to free spectrum space. Thus, the government might have an important role in securing ONdigital's long-term success.

In summary, it can be stated that ONdigital does not achieve a clear differentiation in terms of price or programming. The only significant differentiation is in terms of the means of transmission. Digital terrestrial transmission appears attractive to people who do not want to have a satellite dish or cannot install one for technical reasons. However, those customers may decide to get digital cable to have a larger program offering, so the ONdigital proposition may be attractive to terrestrial viewers who want a simple way of receiving only a few more channels. Current subscriber numbers suggest that this may be a significant market segment. It remains to be seen, however, whether this market segment is sufficiently large to sustain ONdigital's operations in the longer term. It may shrink once more people get accustomed to pay TV and interactive services.

It is likely that ONdigital for many will just form a temporary stage in the move to a larger choice of channels on digital satellite or cable. The challenge for ONdigital would be to delay this "upgrading" of its viewers until more capacity is made available for DTT and then to quickly offer more attractive channels. With the decision to give away decoders, the delaying of viewer upgrading has become impossible. Lock-in (through decoder sales) would have been a strategy to achieve this. ONdigital, however, had to abandon its lock-in strategy due to competitive pressure from SkyDigital. Consequently, with the extension of SkyDigital's offering and new interactive services on digital satellite and cable, it can be expected that the ONdigital customer base will gradually be eroded.

ONdigital had a further significant disadvantage in that it did not have an installed base of pay-TV subscribers. BSkyB has the chance to encourage existing viewers to switch to digital and so establish an installed base more quickly. It is highly unlikely that ONdigital will be able to poach analog BSkyB subscribers. That is why ONdigital is unlikely to ever reach BSkyB's subscriber numbers.

With future potential competition from digital cable, it appears unlikely that ONdigital will be able to compete on equal terms in the long run. The competition for market leadership is likely to be fought out between BSkyB and digital cable. It is feasible that ONdigital will either become a niche operation (although it is not clear what this niche could be) or that it will eventually fall under full control of BSkyB. The latter option would require regulatory approval.

9.4.2.5   U.K. cable: Significant potential

It can be expected that the U.K. market will depict an increasing gap between SkyDigital and ONdigital services. Digital cable is likely to accelerate this development if cable operators devise attractive strategies that can draw in potential DTT customers.

Recent developments in the U.K. cable industry suggest that the digital cable operators may become the strongest competition to SkyDigital. NTL's new strength is its size and the support from new stakeholder France Telecom. This may enable the company to compete on attractive content and to improve on customer service and marketing. Combined with existing competitive advantages such as large bandwidth capacity, full interactivity and TV/telephony packages, this could provide a serious challenge to SkyDigital.

9.4.2.6   A number of niche second movers in Germany

The development of the German market bears few resemblances to the other cases, particularly since second movers have failed to establish any real presence. The MMBG consortium, which established itself as the competitor to DF1, dissolved due to diverging interests and the loss of control over the CA technology. The early digital Premiere package can only really be regarded as a pilot project with a limited geographical coverage, whereby the existing analog channel is simply simulcast on three digital channels (at staggered starting times) plus the provision of four pay-per-view channels. This means that there was nothing really new on Premiere digital. The fact that decoder rental for digital is double as expensive as that for analog does not make it very appealing to switch.

Premiere initially introduced an incompatible decoder (mediabox) and could have started a competing service with the advantage of a base (though relatively small) of analog pay-TV subscribers. Subsequently, however, Kirch took over the majority ownership of Premiere; hence there is at present only one large provider of digital pay-TV services in the German market.

Deutsche Telekom could potentially develop into a serious competitor to Kirch. With MediaVision, the cable operator has started to act as a channel packager, providing pay-TV packages to its cable customers. So far, most of the programming is targeted at niche markets though. As Kirch and Deutsche Telekom are cooperating closely, it is more likely that Deutsche Telekom will remain a niche provider for pay TV, leaving Kirch to serve the mass market. Similarly, several smaller digital ventures, such as RTL World, UPC, or PrimaCom (see Section 7.6), are being established, but those, too, will in the near future only be able to service niche markets.

9.4.2.7 Second movers create healthy competition

It seems that the predictions of the many experts who claimed that there would only be space for one operator in each national (or language) market are not likely to be fulfilled. The French case illustrates this clearly. Although CSN has more subscribers overall, TPS competes effectively, and the competition keeps prices low and services attractive. In the United Kingdom, four services are competing, and competition is proving beneficial for consumers. The above analysis suggests that ONdigital may not succeed in the long run, but competition is likely to continue between BSkyB and the cable industry.

In Germany, the predictions did come true: There is only one digital operator left on the market. However, it does not have an established position in the market. The d-box presently is the only CA system on the German market. However, with a mere 1.6 million users, a viable installed base is far from being achieved. As a result, there is still the possibility of a later entrant that might introduce a different standard to be successful, so long as an improved, differentiated offer could be presented to the consumer.

It is important to note that in each of the studied countries competition is based on different factors: In France, the operators mainly compete on program pricing, in the United Kingdom on distribution media, and in Germany competition is likely to evolve between pay TV and free TV as well as on pay-TV program offerings.

Another crucial finding is that all operators in the analyzed countries abandoned a lock-in strategy through decoder sales. This suggests, first of all, that switching costs, as perceived by consumers, are relatively high. However, it also suggests that the first movers (who make the strategic choice) regard it more important to quickly gain a large subscriber base than to lock viewers in for the long term. Hence, the first movers must be convinced that their competitive advantage in terms of content will be sufficient to retain viewers after the minimum subscription period expires. It further suggests that they expect to be able to match the advantages of their competitors. Canal Plus and BSkyB have already demonstrated this, the former in terms of interactive services, the latter with regard to reduced phone rates and Internet access. It can also be expected that the operators will provide loyalty schemes if churn rates prove to be high. Such loyalty schemes could, for example, be price reductions or bonus channels for long-term subscribers.

9.5 The effects of public debates and regulation

The public debate about regulatory issues and market entry strategies for a new service can be of significance for the adoption of new technology as it can encourage or hinder the credibility of the provider of a new service. It is impossible to precisely measure the impact of such factors, and they might not be decisive factors. Nevertheless, such issues can clearly accelerate or delay market penetration.

A further factor influencing consumer acceptance of digital pay-TV services is the regulation of such services prior to launch. The fact that the regulatory authorities deal with the issues surrounding the launch of digital services and pass legislation so as to prevent anticompetitive behavior can help to create confidence among consumers that they would be protected from monopoly abuse. Furthermore, the existence of a regulatory framework prior to launch of digital services creates certainty for the operators so that they can devise their strategies accordingly.

Television is used by a large majority of the population. A public discussion on TV issues can therefore be expected to receive more attention than other sectors as it is a field with which consumers are familiar and in which they take an interest. Moreover, due to the special characteristics of the media and broadcasting sector (i.e., its impact on public opinion and the fact that it is very specific in terms of regulation), this sector receives attention not only from sector-specific regulators but also from a wide range of consumer organizations, from the government and from the public at large. This wide interest and the direct and visible impact on consumers means that changes and new developments in the broadcasting industry are widely discussed and thus well covered in all kinds of media. For digital TV, this was particularly the case in Germany and in the United Kingdom; it was true to a lesser extent in France.

9.5.1 Media publicity with negative effects in Germany

In Germany, the publicity surrounding digital TV and the main players Kirch and Bertelsmann has had a negative impact on the uptake of digital services. It started with the blocking of the MSG in 1994 by the EU competition authorities on the grounds that this cooperation would be able to control the pay-TV market in Germany. This was the first time that the attention of the public was drawn to the issues concerning potential market control in digital pay TV. Subsequently, there has been an array of changing alliances around Kirch and the MMBG. The Kirch Group with its unpredictable, fast-changing strategic moves has been subject to much speculation and rumor. The discussion of media power that has been raised and fueled by regulators, consumer organizations, and, in particular, the German public broadcasters, concentrates on Kirch as a potential monopolist in the German digital-TV market. The appeals were to prevent Kirch from deploying a proprietary system into which he could lock consumers and thus establish a long-term monopoly position.

Moreover, his heavy investments in attractive film and sports rights as well as the bulk of decoders he ordered from Nokia did not have the desired effect of convincing consumers that a lot of premium content would be made available on DF1 and that the service would surely be a success. Instead, it led to speculation that the Kirch Group was close to bankruptcy. This was supported by the fact that DF1, due to low subscriber numbers, was losing large sums of money and had been looking for international partners that would be prepared to provide large investments.

A further setback, not only for Kirch but also for Bertelsmann and Deutsche Telekom, was the blocking of yet another proposed cooperation of these three by the EU Competition Directorate in June 1998, again on the grounds of potential market foreclosure in German pay TV. This seems to have been the decisive factor for Bertelsmann abandoning its pay-TV activities. For Kirch, this has further damaged his reputation, especially as the rumor spread that his DF1 service might be terminated.

Public awareness of the issues in the German broadcasting sector has also been raised by the ongoing discussion on the regulatory framework and the regulation of new multimedia services. The disputes between the Landesmedienanstalten (state media authorities) and the unclear regulatory responsibilities of the Länder (states) and the federal government also drew much attention in the press. The Landesmedienanstalten have adopted very different positions with regard to the controversial digital pay-TV players. Whereas, for example, the Medienanstalt (state media authority) of Berlin-Brandenburg frequently draws attention to the danger of the media power Kirch might establish, the Bavarian media authority strongly supports the Kirch Group's activities as a cornerstone in the development of new multimedia services and criticizes strongly any activities that could inhibit their free development.

The fact that the German regulatory authorities are divided in their opinions can be expected to create further uncertainty and reluctance on the part of consumers to embrace the new services. Moreover, unlike in the United Kingdom where the government is actively supporting the introduction of digital TV, there has been no official show of support by the administration. The German government did not create significant incentives for companies to launch new digital services, and the regulatory framework, important for providing certainty for investing companies and to create consumer confidence, has been rather vague. Only in 1999 did the government declare its aim to support the move to digital and set the year 2010 as the provisional switch-off date for analog transmissions.

9.5.2 United Kingdom: Public debate and regulation create consumer confidence

The situation in the United Kingdom has evolved differently. Here, too, the discussion around the market power of BSkyB in digital pay TV has been very lively, but it has not had such a damaging effect. This can be attributed to different factors. First of all, BSkyB, though often under scrutiny from regulators in case of anticompetitive behavior, has never actually been convicted, and none of its activities has been blocked by regulators. This has saved the company from gaining an unfavorable reputation with the public at large. It is also strengthened by the fact that Murdoch, being an important player in the U.K. economy, has generally had the goodwill of the government. The reason why BSkyB is largely protected by the U.K. government lies in the power that Murdoch has over U.K. public opinion through his strong position in the print media. Hence, the government cannot afford to seriously attack BSkyB as it would risk vengeance in the form of counterattacks by Murdoch's widely read newspapers. In a sense, there exists an interdependence between Murdoch and the U.K. government, each holding a measure of power over the other, which leads them to treat each other very carefully.

Murdoch has the further advantage of being able to use his presence in the print media to advertise his broadcasting activities and to portray these in a favorable light. This illustrates clearly the specific characteristics of the media and broadcast industries, which due to cross-media ownership can support each other and exert an influence on public opinion.

In the United Kingdom, in contrast to Germany, an extensive regulatory framework was put in place prior to the launch of digital services. The fact that the regulatory authorities deal with the issues surrounding the launch of digital services and provide legislation so as to prevent anticompetitive behavior can help to create confidence among consumers that they would be protected from monopoly abuse. Furthermore, the existence of a regulatory framework prior to launch of digital services creates certainty for the operators so that they can devise their strategies accordingly.

Although the debate on the regulation of pay-TV services has been at least partly responsible for the delay in the launch of such services, it could be argued that this allows more time for the preparation of careful launch strategies and consumer propositions. Although the launch of SkyDigital had been delayed several times, its strategy and product appear clear and targeted, based in particular on both quantity (140 channels) and quality (premium content).

The development and launch of ONdigital was encouraged and enforced through the government support for DTT. Having to compete for a DTT license required the cooperating firms to develop a convincing strategy. The award of the license, which is free of charge to the operators, includes the obligation to launch digital services within one year. This "stick and carrot" approach by the government helped to create a competitor to BSkyB.

9.5.3 France: No need for public debates

In France, the significance of regulatory issues and their public debate has been very limited. The only incentive from the government for the launch of digital TV is provided in the slightly lighter regulation of digital over analog with regard to content requirements. Issues of competition and potential monopoly power in CA systems and the requirement for regulation never received much attention. Considering that the main digital pay-TV operators did not pursue a lock-in strategy anyway, there was no need for such a discussion. French consumers did not have any reason to show reluctance to the new services for fear of lock-in or of finding themselves "stranded" with incompatible technology in the future. This fact is likely to have helped the fast market penetration in France.

9.6 Summary

The above analysis illustrates the influence and interdependence of several factors on the successful launch of digital services in the countries studied. Consequently, the outcomes have been very different.

France, with its favorable existing market structure, was the easiest market to win for digital pay TV. The operators devised appropriate strategies and could avoid extensive public debates on regulation that might have induced consumer reluctance. Furthermore, regulation has been favorable to the introduction of new services, and there were no regulatory disputes that might have discouraged operators from entering the market.

The United Kingdom also had favorable market structures for the fast take-up of digital pay TV, but uncertainty on regulatory issues delayed the launch. This might, however, have been an advantage as it gave companies the time to prepare more carefully their market-entry strategies. At the same time, it raised consumer awareness and reduced uncertainty for both consumers and operators. The fast take-up of digital services since the beginning of 1999 supports this point.

In Germany, all the factors discussed above had an unfavorable influence. With the existing structures, Germany is a very difficult market to develop for pay TV. The market-entry strategies employed by Kirch did not succeed in attracting consumers. His publicity also proved detrimental to subscriber take-up. Finally, government support and an appropriate regulatory framework were lacking.

The analysis of the country cases permits some additional general conclusions. The experience shows that national markets can support more than one digital pay-TV operator, as long as the offerings are differentiated. It appears, however, that the existing pay-TV operators (Canal Plus, BSkyB, and Premiere) succeeded in obtaining the strongest position in digital-TV markets. At present, it is not clear if they are (or will be) dominant. Competition from other digital operators limits their strategic choices. Generally, competition has proven favorable to market penetration, better services, and lower prices for consumers.

Another important finding is that consumer lock-in through decoder sales did not succeed in any of the countries studied; hence, there are no significant first-mover advantages to be exploited with regard to lock-in. This implies that market foreclosure is unlikely and that new market entrants can still be successful if they offer a differentiated service.

Finally, it can be stated that the standardization process for digital-TV technology has provided a favorable basis for effective competition and fast penetration of digital-TV services.

PART

IV

Digital TV in the United States and Japan

Developments for digital TV in the United States

10.1 Introduction

This chapter provides an analysis of the standardization process for digital TV in the United States as well as a market overview of digital-TV penetration for the different transmission media.

Digital satellite TV, for which there is no common standard in the United States, was launched as early as 1994 and has been very successful, counting 14.5 million subscribers at the end of 2000. Cable TV, which has a penetration rate of over 60% of U.S. households, has been slow to launch digital services but is now gaining momentum.

DTT, which is focusing on HDTV, has been fraught with many problems in the United States, including the lack of HDTV content, expensive idTVs and set-top boxes, and an ongoing debate around the suitability of the Advanced Television Standards Committee (ATSC) standard for DTT, called 8 vestigial sideband (8VSB).

10.2 The standardization process

This section analyzes and assesses the U.S. approch to digital-TV standardization as compared with the European process.

10.2.1 Focus on HDTV

An important difference in the standardization approaches of Europe and the United States is the fact that standardization in the United States concerns terrestrial transmission only, whereas in Europe, the initial focus was on DTT but then broadened to encompass all three forms of transmission. The standardization process discussed here applies to terrestrial only, as no cable- and satellite-TV standards have been set.

Another major difference relates to the distinction between HDTV and digital TV. In Europe, there is a clear-cut difference between the development of HDTV, a standardization process that failed in the early 1990s, and digital-TV development, which started in the early 1990s and has since become a major success.

The development of HDTV and digital TV in the United States has proved inseparable. Plans to adopt HDTV as the successor to the American NTSC color standard have been unfolding for more than a decade. In addition to the much improved quality of TV broadcasting offered through HDTV, it was also regarded as crucial to lead the development of this technology over European and Japanese markets. Furthermore, HDTV has been held out as critical to the health of America's consumer electronics industry, which was regarded as saturated. Pressure was thus strong on the U.S. government to take an active role in supporting the development of HDTV.

However, HDTV appeared interesting only for equipment manufacturers—not for broadcasters that did not see any additional revenue coming in from HDTV broadcasts. U.S. terrestrial broadcasters want more channel space to offer a larger variety of channels. This is what they regard as profitable. They have never had much appetite for HDTV, which offers better quality images but uses up more capacity per channel and thus reduces the number of channels that could be made available.

To win the broadcasters for HDTV, the Federal Communications Commission (FCC) decided to give away billions of dollars of valuable spectrum space to broadcasters to enable simulcasting. Each terrestrial broadcaster has been allocated a second 6-MHz channel to simultaneously transmit HDTV and NTSC signals for roughly 10 years. After this transition period, the broadcasters are supposed to return the extra spectrum and owners of analog sets will need to buy HDTV sets or use a set-top box to convert the signals. This arrangement arose from a clever lobbying ploy by broadcasters back in the 1980s: By scaring Congress with the prospect of Japan beating the United States in HDTV development, broadcasters were able to preserve for themselves vacant spectrum space that was in danger of being allocated to non-TV use [1]. This initiative clearly illustrates the dominant interests of broadcasters, and it still prevails today. Broadcasters want to offer multichannel TV, but not HDTV.

10.2.2 A range of proposed DTT standards

In 1987, the FCC established an inquiry, the usual procedure when any important issue is up for decision, led by an advisory committee. The Advisory Committee on Advanced Television Services (ACATS) by law is required to represent all interested parties, including the public. The committee was, however, dominated by industry players with vested interests in the outcome. The aim of the inquiry was to choose an HDTV transmission standard based on performance tests. From 23 proposed standards floated in 1988, six remained by 1991, when the testing was to be started. These six were sponsored by four teams:

- The Japanese Broadcasting Corporation NHK;

- Zenith and AT&T;

- General Instruments (GI) and the Massachusetts Institute of Technology (MIT) (with two proposals);

- Philips, Sarnoff Research Laboratories, NBC, and Thomson (with two proposals).

The long time it took for the regulatory authorities to arrive at a standard turned out to be advantageous for equipment manufacturers, which faced considerable competitive pressures prior to the HDTV sweepstakes in 1990. Indeed, GI entered an all-digital system, a technology that had been regarded as not feasible for fitting into a 6-MHz bandwidth channel. Competitors followed GI's example and within one year all (except NHK) had developed all-digital systems.

10.2.3   The Grand Alliance and the ATSC

In May 1993, after NHK had dropped out, the three remaining teams formed the "Grand Alliance" (GA), merging their technologies and agreeing on cross-licensing. This ended their rivalry in the standards battle.

In May 1993, the GA evolved into a larger committee, including the ATSC. The ATSC had been created in 1982 by a number of engineering societies to be the equivalent of the NTSC that had established the monochrome standard in 1941 and the color-TV standard in 1953. The GA now comprised approximately 120 members from the broadcasting, computer, electronics, and software sectors; industry associations; standards bodies; R&D laboratories; and educational institutions. It also included representatives from Canada and Mexico.

Coming to an agreement within the ATSC, however, proved not an easy task. The ATSC incorporated many proponents of different systems, and nobody was prepared to give up their own format. Moreover, a digital receiver that was already decompressing MPEG-2 contained most of the hardware needed to transcode (any) received scanning format to the scanning format of its display. This provided an opportunity to offer a range of receivers based on the display scanning format. Not all receivers need to have the expensive HDTV scanning capability, but they could still receive HDTV broadcasts. That would, of course, imply an inherent picture quality tradeoff, but that is also reflected in the price.

The ATSC standard chosen for DTT, called 8VSB, finally comprised a total of 14 different scanning formats including many using an interlace scanning format and others using progressive scanning. Interlaced scanning means that an image is scanned (i.e., transmitted) twice. First, the odd, then the even, lines of an image are scanned. At the receiving end, both scanning fields are combined (i.e., displayed sequentially). In effect, the picture repetition rate is doubled, which results in a more fluent picture on the screen. In progressive scanning, in contrast, all scan lines are traced out in every frame.

In early 1996, the standard was ready to be formally mandated by the FCC, but the computer industry vetoed the proposals.

10.2.4   Involvement of the computer industry

The computer industry, which foresaw TV and computer technologies having many interfaces in the future, claimed that it had been shut out from TV technology development and that an old-fashioned TV model would prevail in the ATSC standard. They favored progressive scanning because it avoids interline flickering and is much easier to transcode from one resolution to another. PC monitors all use progressive scanning, mostly 1,024 lines but also 768 lines. A progressive scanning format, 720p, had been developed for TV by MIT and had been adopted by the Zenith/AT&T team. As it already existed and as transcoding to PC-usual resolution was regarded easy, this format was favored by the PC industry for use in TVs.

Broadcasters counterargued that progressive scanning (e.g., the 720p format) was included in the ACATS standard anyway but that the additional inclusion of interlace scanning is important to their live broadcast business—something the computer industry was still not capable of providing. Furthermore, they claimed that progressive scanning can be converted readily to interlaced scanning through a low-cost chip in the receiver [2]. The push for interlace came in particular from certain manufacturers (e.g., Sony) that had already developed 1080i equipment and wanted to sell this before shifting to progressive.

The dispute could not be resolved. Interestingly, the computer industry, though very influential in the United States, was unable to assert itself against the broadcast industry. This may be due to the fact that it entered the digital-TV standardization process late, only in the mid 1990s. At that time, the process was already advanced, and the involved players, especially the FCC, were keen to formally set a standard soon.

To quickly move to formal standard setting, the FCC adopted the compromise of allowing a number of different systems to be deployed. The FCC Digital TV Standards Decision, released on January 29, 1997, does not include requirements with respect to scanning formats, aspect ratios, and lines of resolution. This implies that broadcasters can use incompatible systems for digital transmission. This may result in slowing down consumer acceptance of DTT and delaying the switch-off of analog transmission that is planned for 2010.

Already, the main terrestrial program providers in the United States, called the broadcast networks, are divided between different scanning formats. ABC/Disney and Fox aim to incorporate complementary PC-type functions. Both adopted the 720p format. NBC and CBS are opting for higher picture resolution and chose the 1080i format.

10.2.5 Assessment of the standardization process in the United States

In April 1997, the FCC ruled that the four largest Networks (ABC, CBS, NBC, and Fox) would have to start digital broadcasts in their "top 10" (i.e., largest) markets by May 1, 1999. Affiliates of these networks in markets 11 to 30 must be on air with digital by November 1, 1999, and all other terrestrial broadcasters by May 2002. These requirements have not been altered, but many of the smaller broadcasters may not be able to fulfill them on time.

The FCC did, however, make concessions with regard to HDTV in that broadcasters are not required to transmit high-definition programming or simulcast their analog programming on the digital channel. This is a victory for the broadcasters, which can now use their additional capacity for multichannel TV. With this opportunity, many networks even started digital broadcasting ahead of schedule, as early as November 1998.

This decision is likely to mean a delay in the penetration of HDTV sets, especially as cable and satellite broadcasters have no plans to provide large-scale high-definition programming. The cable operator TCI even uses half-resolution images, providing VHS-quality pictures that are only half as clear as the original. This enables the operator to transmit more channels on the available bandwidth.

The developments in the United States illustrate that a chain is only as strong as its weakest link. The pace of adoption can be set by those players that are least interested in the new standard. It also shows that consumer demand is the critical determinant in the adoption of new technologies. The broadcasters believe strongly that they are more likely to gain revenues from multichannel TV and data-casting than from high-definition images. In Europe, this had been recognized after the failure of the HD-MAC initiative, and digital-TV technology was developed with the later possibility of high-definition transmission, but with an initial focus on multichannel TV. In this way, the broadcasters could be attracted to joining the DVB standardization process with the prospect of being able to offer more content over their networks.

It is the author's opinion that the acceptance of a standard and the launch of DTT could have happened much earlier had the focus in the United States not been so strongly on high-definition, but rather on digital as a way to use transmission capacity more efficiently. As will be shown in Section 10.3, the penetration of other digital transmission platforms proves that there is broadcaster interest and consumer demand for multichannel pay TV, but not so much for high definition.

10.3   Overview of the U.S. television market

The U.S. market is fundamentally different from the European market. First of all, viewing times are much higher in the United States than in

Europe. Average daily TV household viewing with nearly 7.5 hours is around double that in Europe. Of all households, 67% own two or more receivers. Second, the United States has the largest advertising market in the world, 30% larger than that of the whole of Europe. Third, the United States is a true "multichannel" country: Over 60% of TV households receive 30 to 100 channels via cable TV. Finally, the United States does not have a dual structure of broadcasting: There are no public broadcasters financed through license fees, but all networks are financed either through advertising or subscription payments. This is even true for the Public Broadcasting Service (PBS), which is one of the large broadcast networks.

The program providers in the United States can be broadly distinguished into four categories. The main terrestrial broadcasters are the so-called broadcast networks: CBS, NBC, ABC, PBS, Fox, UPN, and Warner Brothers. "Stations" are regional or local broadcasters. There are around 1,800 in the United States. Some are owned by the networks; some are affiliated with the networks; and again others are independent. For cable transmission, there are the basic cable networks (e.g., MTV, ESPN, CNN, USA, TNT, and Lifetime), which offer basic cable packages. A third category are the pay or premium cable networks (e.g., HBO, Showtime, and Cinemax). Finally, pay-per-view offerings are supplied by Don King and SNI to cable and satellite. Terrestrial transmission has no pay-per-view [3].

Figure 10.1 illustrates that cable, with a penetration rate of more than 60%, is the most popular transmission method. Satellite has only recently developed but shows strong growth and is stealing market share in particular from cable.

10.4 Digital satellite TV

The United States saw a number of competing digital satellite operators aiming to win over unsatisfied cable and terrestrial viewers. This section gives an overview of the players and their developments.

10.4.1 Digital spurs the emergence of satellite TV

Prior to digital offerings, there was no significant satellite TV in the United States. The market is characterized by regional and local programs with only a few hours per day devoted to nationwide programs. Consequently, there was no strong demand for direct broadcast satellite services via analog transmission, since these would not have had the capacity

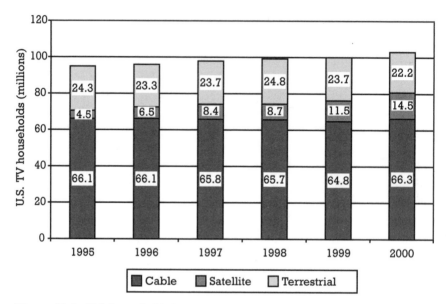

Figure 10.1   U.S. households by transmission medium, 1995–2000.

to broadcast all the regional and local varieties. However, when digital transmission became available with the possibility to cheaply transmit large numbers of channels via satellite, this distribution medium developed rapidly. The main drivers in the development of digital satellite TV have been sports events and pay-per-view movies. Digital satellite permits the transmission of sports events across the whole of the country. The capacity for many channels implies that a wide choice of pay-per-view content can be offered.

An important factor in the fast market penetration of digital satellite TV is the widespread dissatisfaction of consumers with the cable operators. In particular, cable operators are widely criticized for their poor customer service. Furthermore, the United States is a very large country, and cable does not pass households in many rural areas. In those regions satellite subscribers grew fastest in the mid- to late 1990s.

During that time, a disadvantage for digital satellite, despite offers of up to 200 channels, was the fact that none of the local channels of the large networks ABC, CBS, NBC, Fox, or PBS was available via satellite. Since 2000, digital satellite services are now increasingly broadcasting local stations as well. As a result, U.S. satellite services are now gaining more and more subscribers in cabled areas as well.

From 1994, DirecTV, PrimeStar, USSB, Alphastar, and Echostar entered the satellite market with digital offerings, and subscriber growth rates have been rapid (see Table 10.1). The first digital satellite offering, DirecTV, acquired one million subscribers only 13 months after its launch in June 1994.

This is quite an achievement, especially in light of the fact that no standards had been set for digital satellite TV. The operators use incompatible transmission technologies and set-top boxes. DirecTV employs its DSS transmission technology; PrimeStar developed the Digicypher system; and Echostar adopted the European DVB-S standard.

By early 1999, the digital satellite market had consolidated, leaving two digital satellite operators, DirecTV and Echostar, competing for subscribers. The following sections give an overview of the developments.

10.4.2 DirecTV: The market leader

DirecTV was one of the first providers of digital satellite TV in the United States. It was launched in June 1994, the same time that PrimeStar launched a competing service. The program offering comprises around 200 channels, and many of these are pay-per-view channels. DirecTV achieved market leadership especially through exclusive program packages consisting of popular sports events, and through pay-per-view.

To receive DirecTV, viewers are required to buy a satellite dish and a set-top box, though at subsidized prices. Total installation has often been available for $200 or less.

10.4.3 United States Satellite Broadcasting Company: A niche operator swallowed up

The United States Satellite Broadcasting Company (USSB) is a subsidiary of Hubbard Broadcasting, the company that first introduced U.S. direct

Table 10.1
Subscribers to U.S. Digital Satellite Services, 1997–2000 (millions)

	November 1997	November 1998	End of 1999	End of 2000
DirecTV (incl. USSB)	2.99	4.17	8.08*	9.5
PrimeStar	1.88	2.21	—	—
Echostar	1	1.71	3.41	5
Alphastar	0.055	—	—	—
Total	5.925	8.09	11.49	14.5

* Now including Primestar subscribers.

satellite. Together with Hughes Electronics, Hubbard, in the early 1990s, developed the digital satellite system DSS. USSB subsequently acquired five transponders from DirecTV and uses the same transmission technology and set-top box.

Launched in June 1995, USSB only offered 25 channels with a focus on recent entertainment films. Despite this very small channel offering, it gained well over one million subscribers over around two years. The advantage for USSB was that it deployed the same technology as DirecTV and was able to evolve into a niche supplier. Since June 1997, the USSB offer was marketed together with DirecTV and the companies have since merged.

10.4.4 PrimeStar: Another competitor swallowed up

PrimeStar, the second largest digital satellite broadcaster, is a cooperative effort of several cable operators and broadcasters. It is owned by News Corp. (30%), TCI (14%), Cox Cable Communications (14%), Continental Cable Vision (14%), Time Warner (14%), and Comcast Cable Communications (14%).

The program offer is similar to DirecTV, but satellite dishes and set-top boxes can be rented and the packages are cheaper. Rental and subscription costs are combined in a monthly fee, with the smallest program offering costing "one dollar per day." By the end of 1998, Prime-Star had around 2.2 million subscribers, around half as many as DirecTV/USSB.

In the beginning of 1999, DirecTV announced the acquisition of PrimeStar. Within two years, the PrimeStar customers have been converted to DirecTV, which deploys a different set-top box that needs to be purchased. As an encouragement for subscribers, the company offered an equipment leasing arrangement for PrimeStar subscribers.

10.4.5 Alphastar: An unsuccessful niche operator

Alphastar is a subsidiary of the Canadian Tee-Comm Electronics. In July 1996, it launched a digital satellite offering comprising 100 TV channels and 30 radio stations, aimed particularly at viewers in Canada and the Caribbean. By the end of 1997, the service had gained around 55,000 subscribers. About two years after launch, the service was terminated due to unprofitability. Alphastar relaunched in May 1999 as a satellite Webcasting network, providing multicast streaming of video and audio to the Internet.

10.4.6 Echostar: A latecomer catching up

Echostar's DISH Network was launched in April 1996, almost two years behind the first digital satellite operators. Nevertheless, it has quickly been gaining subscribers, passing the 1 million mark even faster than DirecTV, after only one year. Since then, the company has been growing steadily: By the end of 2000, Echostar had over 5 million subscribers. The popularity of the service is attributed to low prices for programming combined with good customer service.

Echostar initially offered around 130 channels. With the transfer of satellite and broadcast assets from News Corp. and MCI's failed digital satellite venture ASkyB in 1997, Echostar took control of more satellites, giving it a channel capacity of over 500, the largest number of digital-TV transponders in the world. This has enabled the company to become the first satellite broadcaster to offer local channels, seen as essential if satellite is to compete effectively with the cable operators.

To quickly gain subscribers, Echostar has been offering potential clients a free digital satellite system (worth $249) if they subscribe for one year ($48.98 per month). This offer provides them with 100 video and audio channels including two premium film channels. Gradually, the broadcaster has introduced interactive television services and high-speed Internet access. Services include on-demand interactive multimedia, software downloading, games, electronic publications, and on-line shopping.

10.5 Digital cable TV

Digital cable rollout in the United States had a slow start in comparison to satellite. This is mainly due to the fact that for many cable systems digital transmission requires extensive rebuilding. By the end of 1997, only around 10,000 households could receive digital cable TV. Since then, cable operators have gradually been offering more and more digital services over their networks and are now acquiring subscribers at a faster pace.

The main advantage of cable transmission has been that the programs of the large national networks can be received. As noted earlier, those have initially not been available over digital satellite. However, there have been disputes as to the "must-carry" requirements of the networks' planned HDTV programming. Under existing must-carry rules, the cable operators must transmit all terrestrial channels within their normal coverage areas. They are, however, reluctant to use up their capacity for bandwidth-intensive HDTV transmissions but would rather offer a larger variety of standard-definition channels instead. Some operators claim they would like to charge a higher fee for high-definition transmission as those use up more capacity.

In January 2001, the FCC took a partial decision on must-carry rules for cable operators. The commission decided that cable operators should not be forced to carry terrestrial broadcasters' analog and digital signals simultaneously. However, the FCC also requested more information before making a final decision. The issue of HDTV programming is still outstanding.

Among the larger cable operators offering digital services are AT&T BIS (formerly TCI), claiming 2 million subscribers, Time Warner (613,000 subscribers), Comcast (515,000 subscribers), and Cox Communications (450,000 subscribers) as of mid-2000. Package prices start from $6 per month. For all cable operators, pay-per-view is the main driver for digital TV in terms of revenues generated. Pay-per-view has already been available (and profitable) on analog cable; however, digital can provide more channels and hence a larger choice for viewers.

By the end of 2000, around 7 million U.S. households received digital cable TV services, up from 3.6 million in 1999. This is, however, still only 10% of all cable households, or 6.8% of total TV households.

The cable operators have been slow in jumping on the digital bandwagon, but are now making reasonable progress. With the strong competition from digital satellite, they are losing some of the high market share they had in analog TV. To stay competitive in the long run, they will have to make use of the distinctive advantages of cable (i.e., offer interactive services or attractive packages combining TV with telephony, Internet, or other data services).

The FCC has announced its intention to mandate a common, open standard for digital cable TV. In light of the fact that most cable operators have already launched services, this initiative may cause disputes if some cable operators have to give up their already employed technologies.

10.6 DTT: Driven by the regulator

This section describes the emergence of digital terrestrial services in the United States, explaining the divergence between government objectives and broadcasters' demands.

10.6.1 Government incentives

As noted in the beginning of this chapter, the regulatory focus in the United States has been on digital terrestrial transmission and the role of HDTV in this context. To encourage the development of digital HDTV, the FCC decided to transfer the existing analog services to a digital environment. This was done by assigning frequency spectrum to each of the existing terrestrial broadcasters at no extra charge. The decision has proved controversial, with a number of industry players believing that these frequencies should have been auctioned. There is considerable pressure for reuse of the frequencies, and the FCC has adopted a policy of "use it or lose it." This is similar to the U.K. government's approach to provide free additional spectrum space to existing broadcasters under the obligation that they start digital services within a certain time period (see Section 8.3).

The FCC mandated that the largest broadcast networks (ABC, CBS, NBC, and Fox) have to start DTT services in the 10 largest markets (certain metropolitan areas) by May 1999, in the top 30 markets by November 1999. All other commercial stations have to begin no later than May 2002 and noncommercial stations one year later. Many stations switched on digital signals ahead of schedule. The FCC plans to terminate analog services by May 2006. To further encourage the rapid transition to DTT, the U.S. federal budget for the fiscal year 2000 included a proposal to introduce a lease fee of $750,000 annually for analog spectrum used by the commercial TV stations.

By February 2001, 182 stations had launched digital broadcasts, but another 1,100 stations have yet to begin. The FCC is already discussing an extension to the 2006 switch-off date and the demand for broadcasters to be charged for analog spectrum use beyond 2006 has been renewed.

10.6.2 DTT take-up is slow

By May 1999, only 35,000 DTT sets had been sold in the United States. This is very little in comparison to 20 million total U.S. TV-set sales in 1998. By the end of 2000, DTT-set sales were still below 100,000, making analog switch-off in 2006 an unlikely scenario. The slow take-up is attributable to the high price of HDTV sets. They cost between $1,000 and $14,000 whereas standard TV sets cost between $100 and $700.

Similarly, DTT set-top-box sales have been disappointing. By the end of 1999, only 64,000 households had acquired such a box. With a retail price at around $1,500, this is costly for most consumers. Furthermore, most of the HDTV sets on the market additionally require a set-top box to receive HDTV programming. It is acknowledged that HDTV equipment, at this point in time, is mainly aimed at professional markets (sports bars, etc.) and high-end consumer markets. Industry players claim that further reasons for the slow penetration of DTT are the lack of HDTV content, problems with the transmission standard, and the lack of mandated cable carriage.

To boost sales of HDTV sets, consumer electronics manufacturers have started to sponsor HDTV programming production. Mitsubishi sponsored CBS for one year to make its prime-time programming in high-definition. Similarly, Panasonic provides HDTV production equipment for ABC and sponsors HDTV programming for CBS. There are, however, no plans to subsidize costs for HDTV sets for the consumer. Consumer research in 1998 by the Yankee Group found that 86% of surveyed households are "very or somewhat satisfied" with standard-definition TV picture quality and reception. The study does not reveal if the surveyed individuals had actually seen HDTV, which might have changed their level of satisfaction. However, it is still unlikely that the perceived added value from higher image quality would be sufficiently high to justify investment in a costly HDTV set.

10.6.3 Debate around the suitability of the 8VSB standard

Already in 1999, criticism arose in the United States about the technical efficiency and suitability of the 8VSB standard. The Sinclair Broadcasting Group, a large program provider, carried out extensive tests and claimed that the 8VSB standard was inferior to the DVB-T OFDM standard with regard to picking up signals using simple indoor and outdoor aerials. The tests carried out revealed that indoor antennas could only receive two out of five DTT channels at a time. Sinclair filed a petition with the FCC in October 1999 to allow U.S. broadcasters to choose either the OFDM standard or 8VSB as a modulation system. Shortly after, the Consumer Electronics Association (CEA) filed a response strongly opposing the petition and defending 8VSB.

The consumer electronics manufacturer Zenith also defended the U.S. standard arguing that the European and the U.S. systems were optimized for different conditions (e.g., high-power UHF and VHF transmitters in the United States and low-power single frequency networks in the United Kingdom). Zenith argues that all such technical decisions involve performance tradeoffs. Zenith demonstrated 8VSB superiority over OFDM for a number of noise interference conditions.

During the course of 2000, the debate continued with more and more players picking sides. The CEA, Zenith, and Nxtwave are the main defenders of 8VSB. Among the OFDM proponents are Sinclair, NBC, Pace, Nokia, and Disney/ABC. In January 2001, having carried out a detailed evaluation, the FCC reconfirmed exclusive use of 8VSB for U.S. DTT. At the same time, the ATSC issued a request for proposal for 8VSB enhancements, in particular with regard to portable and mobile reception.

The FCC decision reflects not only the technical issues around the standard but also potential commercial problems should two incompatible standards be permitted in the United States. It is generally acknowledged that a single standard for a single market is favorable to enable fast penetration. Consumer confusion over standards could inhibit take-up of DTT. With 8VSB already employed and major investment having gone into equipment manufacturing, it appears favorable to keep 8VSB and rather invest more in its improvement. Hence, at this point in time, the FCC took the right decision in reconfirming exclusive use of 8VSB in the United States. The ongoing debate, however, illustrates that the United States may not have picked the best available technology in the first place.

10.7 Summary

The United States was the first country to introduce digital TV. Multichannel services via satellite commenced as early as 1994. Over a period of six years, the country has reached a digital-TV penetration level of 21%. Around 21.7 million households received digital-TV services by the end of 2000. Figure 10.2 shows that the bulk of that is received by satellite. However, digital cable shows strong growth rates now. DTT, in contrast, depicts a slow uptake. This can be seen as a direct result of the standardization process.

10.7.1 Standardization focuses on costly HDTV-set sales

The standardization process in the United States has been driven by the set manufacturers, which hope for large revenues from a rapid adoption of digital TV. Under their pressure, and in the fear of getting behind Japan and Europe in high technology, the U.S. government created incentives for broadcasters by offering them free spectrum space for digital broadcasts. The additional spectrum space made available is a temporary "gift"

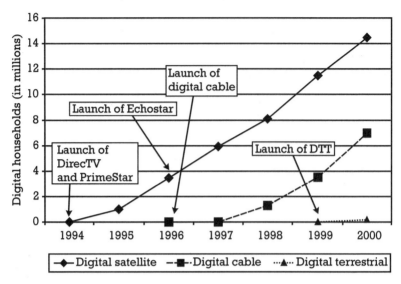

Figure 10.2   Digital-TV take-up in the United States.

to enable a smooth transition and to ease the burden of switching costs from analog to digital.

Standardization has attempted to promote the HDTV format for terrestrial transmission. Satellite and cable were excluded from the standardization process, resulting in the deployment today of three incompatible digital satellite systems. For cable, too, no standard has been set, but the FCC decided that cable set-top boxes must be made interoperable by 2000. Despite this suboptimal standardization outcome, digital satellite and cable have been very successful in the United States.

For DTT development, the focus has been on achieving a fast penetration of HDTV sets, which is of course the primary interest of consumer electronic manufacturers, which have much to gain from sales of such sets. This is in contrast to Europe, where emphasis has been on the penetration of set-top boxes as converters and where idTVs are only gradually being introduced. This is a strategy useful to avoid excessive switching costs for consumers. In the United States, no attempt has been made by the involved actors to reduce, at least initially, the costs for consumers, for example by subsidizing HDTV-set costs.

The standardization approach has evolved through a committee process. As discussed in Section 3.3.3, such a process holds the danger of being influenced by participants with strong vested interests. As in the DVB process, the vested interests of certain industries prevailed in certain aspects of the standard. In Europe, the dispute arose over CA; in the United States, over high-definition and later the scanning format for transmission.

The standardization process in the United States has initially been dominated by the influence of consumer electronics manufacturers pushing for HDTV. The government has attempted to secure the support of broadcasters, but with limited success. In the debate around scanning formats, the broadcasters could bring their influence to bear against the computer industry. It is possible that the FCC supported the broadcasters against the powerful computer firms so as to ensure the broadcasters' further cooperation in the development and launch of DTT.

In contrast to DVB, in the United States, consumer demands have not been recognized, particularly with regard to high-definition images. These are not seen to merit a major investment in new TV sets. Consumers are prepared to invest in new equipment and pay for a larger variety of channels; however, HDTV sets are unpopular due to their high prices and the lack of interest in high-definition images. A 1999 report by the consultancy Strategy Analytics estimates that only 6.7% of U.S. households will have HDTV sets by 2005, whereas 63% would have digital set-top boxes.

10.7.2 Comparison of European and U.S. standardization processes

Traditionally, comparisons have been made between the standardization processes for HDTV in the United States and Europe, with a focus on the distinction between the centralized approach in the EU and the market-oriented approach in the United States. Europe's HDTV approach is widely recognized as a failed interventionist standardization policy whereas the U.S. approach was regarded a success because it had led to the development of an all-digital technology. However, 1993 is not the end of the story. By then, there was still no digital standard in sight in the United States. For a true comparison and assessment, the horizon needs to be widened to incorporate the further developments in the United States and the DVB process in Europe. The DVB Group agreed on common specifications for cable, satellite, terrestrial, and microwave transmission; service information; and other factors within a period of three years. The United States took 10 years to agree on a standard that still allows incompatible transmission formats, and the ATSC standard is for DTT only. The United States has no common standards for digital satellite and cable TV. At present, the satellite operators use their own standards, and there is no national agreement on EPGs, CA, or common interface systems. Such decisions have been left entirely to the market.

Despite there being several incompatible transmission systems on the market, satellite broadcasters have succeeded in attracting large numbers of viewers to subscribe to their services. Cable operators are now following suit. This illustrates that there is demand for multichannel pay TV. However, consumers appear reluctant to acquire costly HDTV sets as that would mean being locked into a certain system, incompatible to others.

10.7.3 International acceptance of DVB and ATSC standards

A further success of DVB over ATSC regards the penetration of digital-TV standards in countries outside Europe and the United States. The ATSC standard for DTT has so far only been adopted in Canada and Mexico. DVB has been more successful: The satellite standard DVB-S is being employed in several countries in Asia-Pacific, Africa, the Middle East, and North and South America. DVB-T, the terrestrial specification, was chosen for Australia, New Zealand, India, Singapore, and Israel. Taiwan, Korea, and Argentina initially planned to adopt the U.S. DTT standard, but are now reviewing their decision and are conducting side-by-side tests of DVB-T and 8VSB. It will probably be those countries that can decide, free from national technology policy issues, which standard is more suitable—and the outcome is likely to depend on the prevailing conditions and requirements of the players and may therefore not be the same for each country.

Endnotes

[1]   Shapiro, C., and H. Varian, *Information Rules, A Strategic Guide to the Network Economy*, Boston, MA: Harvard Business School Press, 1998, p. 219 f.

[2]   See de Bruin, R., and J. Smits, *Digital Video Broadcasting: Technology, Standards and Regulations*, Norwood, MA: Artech House, 1999, p. 65.

[3]   Bird, A., "Finanzfluss im US-amerikanischen Fernsehsystem," *Tendenz Magazin für Funk und Fernsehen der Bayerischen Landeszentrale für neue Medien*, No. IV, 1998, München, p. 28 f.

Digital TV in Japan: From pioneer to laggard?

11.1 Introduction

The development of digital-TV services in Japan has been hampered by the legacy of the analog HiVision standard. As in the United States, standardization focused on DTT only. Digital terrestrial services will not commence until 2003.

Digital satellite was launched in 1996, using the DVB-S standard. The peculiar structure of the satellite sector, being divided between broadcast satellites (BSs) and communications satellites (CSs) has prevented rapid take-up. By the end of 2000, only around 2.5 million households received digital satellite services.

Cable TV has developed very slowly in Japan and is characterized by a highly inefficient, fragmented market. Cable operators have been late to develop digital services. They mostly benefit from digital broadcast satellite feeds into cable head-ends.

11.2   The legacy of HiVision

Japan was the first country to start the development of a new advanced TV technology after color TV had been widely adopted and the market for TV sets showed signs of saturation. As early as the 1970s, the Japanese government had coordinated and subsidized the development of various technologies needed to make HDTV work.

The only public broadcaster, NHK, had determined a new standard (HiVision) for HDTV by 1984. Because a high-definition picture requires a higher bandwidth than a normal terrestrial channel, the Japanese decided that satellite transmission was the only option available, and invented the transmission technology multiple sub-Nyquist sampling encoding (MUSE), which compressed the original signal sufficiently to fit into the bandwidth available on a single satellite channel. MUSE was developed with the aim of compressing and delivering a higher resolution image by satellite in the early 1970s. Japanese firms and the government together invested as much as ¥153 billion ($1.3 billion[1]) to develop HDTV technology.

Having chosen their standard, the Japanese government in 1986 proposed its HiVision standard to the international standards body, the CCIR, as an international standard for the production of HDTV material in the studio. Canada and the United States initially supported the Japanese position. The European governments, in contrast, decided to block the proposal. This may have prevented the Japanese technology from becoming a world standard. Once the Europeans had rejected it, the United States, too, started to develop its own.

Japan, meanwhile, went ahead implementing its standard for the Japanese market. By 1991, NHK broadcast eight hours per day in the HDTV format. TV sets, however, remained extremely expensive and the advantages of high-definition were evident only on the largest sets (36 inches and more). By 1994, the cheapest HDTV sets still cost ¥700,000 ($6,000) and only about 20,000 had been sold in Japan. Since then, sales have accelerated, but they are still far from achieving mass-market penetration despite two-thirds of airtime being broadcast in HDTV format. By the end of 1999, a total of only 820,000 analog HDTV sets had been sold. In total, there were around 1.96 million HiVision homes if those equipped with standard TV sets and HiVision-to-NTSC converters are included.

---

1. For the entire chapter, the exchange rate of $1 = ¥117.5 is adopted.

The timing of the introduction of the new TV standard proved criti-cal. In an attempt to be first on the market insufficient attention was paid to the high switching costs. The HiVision standard did not offer a suffi-cient improvement over the existing system to justify the high switching costs. Moreover, in this field of rapid technological change it quickly became apparent that the hybrid analog/digital standard would soon be obsolete.

Moving too early meant making compromises in technology and entering the market without sufficient allies. NHK tried to go it alone in the early 1990s with its own HDTV system, with costly consequences: Not only was NHK's MUSE system met with consumer resistance in Japan, it also left the Japanese behind the United States and Europe in the development and implementation of fully digital TV.

11.3 The government plan for DTT

The Japanese government has traditionally played an important role in directing technology-driven innovations. The Ministry of Post and Tele-communications (MPT) is responsible for the industrial policy in the tele-communications and broadcasting sectors. Technology development is driven (and financed) mainly by companies; however, the government acts as a supporter and influences shaping industries.

11.3.1 Development of the ISDB-T standard

The rapid innovation and implementation of digital compression technol-ogy in the United States and Europe came as a great shock to the MPT and Japan's broadcasters, because they were forced to realize that they had become latecomers in digital innovation. Although Japan was slow to switch to digital broadcasting, the country is now showing rapid improvement in areas such as the development of mobile broadcasting services, a potentially important new market segment.

Japan's broadcasting policy changed drastically after 1994 when digi-tal broadcasting gained worldwide attention. Aggressive U.S. and Euro-pean moves to introduce digital technology gave a strong incentive to Japan to join the digital television market.

Japan officially started the development of digital TV in 1994. The MPT founded the Digital Broadcasting Development Office to coordinate the development. By that time, the European offices of several Japan-based companies had already participated in the DVB Group. That is likely to be the reason why Japan also chose the MPEG-2 compression standard. Furthermore, the Japanese proposal for digital transmission systems is very similar to the DVB specifications. For DTT, Japan developed its own standard: the integrated services digital broadcasting (ISDB) standard. This has not proven to be a successful export but it is speculated that ISDB and the European DVB-T technology could merge as the terrestrial transmission chip sets (OFDM) used are almost identical. This raises the question why Japan did not adopt the DVB-T standard as it is, a move that could have proved advantageous for Japanese consumer electronics firms. One likely reason is that DVB-T was not finished at that time and Japan probably did not want to wait as it was "late" already in the development process. Another reason might stem from industrial policy objectives—the desire to retain a "homemade" technology.

11.3.2 Government plans for DTT penetration

The MPT changed its tune and began promoting digital innovation, not only via satellite but also via terrestrial transmission. However, since conversion to digital television will be costly, the MPT is now required to balance the desire to quickly advance technological development while ensuring the support of industry players.

In support of the fast penetration of this technology, the MPT announced its digital terrestrial broadcasting plan in June 1998. Initially, terrestrial broadcasters were expected to introduce digital technology in Tokyo by 2000, and then other major metropolitan areas such as Osaka and Nagoya by 2003. Commercial broadcasters showed extremely strong resistance to these plans due to the large investments needed for digitization. The revised plan now foresees DTT launch by 2003.

A detailed cost analysis by the MPT, NHK, and commercial broadcasters reveals the high transition costs. These costs arise from the need to reshuffle analog frequencies when DTT is launched. The cost of analog simulcasting, including new receiver equipment for households, is estimated at a total of ¥85.2 billion ($725 million). Around 2.5 million households will need new aerial and other receiver equipment to maintain analog reception, costing ¥54 billion. TV stations will need to invest ¥31.2 billion for new analog transmission equipment.

Many of the small and local broadcasters cannot afford this investment. Others fear that after making the investment they may not be able to compete with the large commercial stations.

The MPT proposed in mid-2000 that the Japanese government should not only subsidize the transition costs in the switch to DTT but shoulder the total cost. In return for this support, the MPT expects broadcasters to return unused spectrum to the government once the transition is completed in 2010. The government could then auction these frequencies for mobile or other services. To encourage a swift transition period, the government announced that analog broadcasting would be terminated in 2010, when the MPT expects digitization of terrestrial, satellite, and cable TV to be completed.

While the government wants to promote digital terrestrial broadcasting to restore Japan's position in global digital markets, commercial broadcasters as well as NHK are reluctant to go digital in terrestrial broadcasting. This is due to the large costs for DTT, as shown above, but also to the fact that they will have to introduce both digital terrestrial and satellite broadcasting at almost the same time. Section 11.3 looks at the satellite sector.

11.4   Satellite TV: Nothing new to the Japanese viewer

Satellite TV had existed in Japan since 1984. This section looks at existing analog offers as well as the new digital services.

11.4.1   Broadcast and communications satellite TV

Until 2000, satellite TV in Japan was characterized by two incompatible types, BSs and CSs. The former are high-powered satellites with a high bandwidth whereas the latter provide a lower bandwidth and are low-to-medium-powered.

11.4.1.1   BS services

Satellite TV in Japan started with the launch of a BS by NHK in 1984. This symbolized the new era of broadcasting technology, which expanded the technological frontier from solely terrestrial broadcasting to both terrestrial and satellite broadcasting. Moreover, satellite broadcasting systems made possible encrypted services, which spread the idea of charging viewers for programming. In the late 1980s, the broadcasting law and the radio law were amended to develop a new financial system applicable to pay-TV services.

Via BS, NHK provides two public service channels plus a third channel devoted to HDTV broadcasting. NHK has by far the largest number of satellite viewers. From a total of 13 million subscribers to BS services in

1999, 11 million watched NHK programming. This represents a quarter of all TV households in Japan. The public broadcaster is highly popular for quality content and offers a range of exclusive sports and cultural programs. The subscription fee system has provided a favorable situation for NHK through a stable income. A terrestrial color-TV license costs ¥1,370 per month; a satellite license (including terrestrial) is more expensive at ¥2,300. The differing price results from the fact that the NHK channels provided on satellite are not the same as those via terrestrial.

The remaining 2 million BS subscibers were subscribers to WOWOW, the only commercial channel transmitting via a broadcast satellite. This pay-TV service, which started in 1991, is devoted primarily to films and thus provides a complement to the general-interest NHK services.

11.4.1.2 CS services

Broadcasting through CSs started in 1985. The liberalization of the telecommunications market and privatization of a public telegraph and telephone company (Nippon Telegraph and Telephone Public Corporation, now NTT) had a strong influence on the market structure of Japan's broadcasting sector. For the first time, it allowed a telecommunications company to enter the broadcasting business through CSs. With CSs being used for both telecommunications and broadcasting services, this started the convergence process, eliminating the market boundaries between the two sectors, much earlier than in Europe or the United States.

After the relaxation of entry regulations into satellite broadcasting by communications satellite in 1989, two satellite telecommunications carriers launched analog broadcasting services in 1992. JCSAT-2, operated by the CS Baan Group, offered 4 TV and 14 radio channels; Superbird-B, run by Skyport, supplied 9 TV channels. The analog CS broadcasters, however, are much less popular than the NHK BS service. By 1997, they only had a total of 150,000 subscribers.

11.4.1.3 Merging CS and BS

A major problem for the CS operators is the incompatibility of CS and BS equipment. Although their programming may be attractive, the majority of viewers want to be able to watch the public broadcaster's channels and maybe some additional channels. With incompatible equipment, in a purchasing decision, priority is likely to be given to the NHK satellite dish—even more so since the BS dishes are smaller and cheaper. Moreover, most TV sets have BS decoders built in, whereas CS decoders need to be purchased separately.

The analog program offerings via CS have so far been insufficient to provide a real differentiation to broadcast satellites. Due to their lower bandwidth, Css can only offer a limited number of channels. Digital transmission, however, allows for the provision of a much larger number of channels. This is what SkyPerfecTV and DirecTV Japan are aiming for.

To improve the competitiveness of CS broadcasters, the MPT induced a merger of broadcast and communications satellite platforms. The current BS-4a satellite has been phased out since December 2000, when a new digital broadcast satellite, BS-4b, was launched. The new satellite carries programming from eight broadcasters: NHK, WOWOW, Star (showing movies only), and five terrestrial broadcasters. A new communications satellite is to be launched into the same orbital position as the broadcast satellites to allow BS viewers to also receive CS digital channels via existing dishes.

Meanwhile, Matsushita, Sony, and Toshiba are developing set-top boxes and idTV sets that will be able to receive programming from both CS and BS broadcasters. The first products are expected by the end of 2001.

11.4.1.4 Digital CS offerings

Digital satellite broadcasting is the immediate challenge in Japan's broadcasting environment. Two broadcasting service operators, SkyPerfecTV and DirecTV Japan, using CSs, have been competing head-to-head in the multichannel digital pay-TV market. The two digital CS operators used incompatible decoder systems, although a dual-band Sony receiver is available. No standard has been set for digital satellite transmission, and the operators both chose the European DVB-S technology.

PerfecTV was first to launch digital satellite broadcasting in June 1996. In May 1998 it merged with Japan Sky Broadcasting to create SkyPerfecTV. The aim was to seek jointly a dominant position in the market. The five main shareholders of SkyPerfecTV with 11.4% each are Rupert Murdoch's News Corp., Fuji Television Network, Sony, the trading house Itochu, and Softbank. One of the advantages of SkyPerfecTV is that its member companies can provide various categories of programming. For example, Sony can offer films and music, and Fuji Television can provide news, sports, and dramas.

SkyPerfecTV offers 194 TV channels and 106 digital audio channels, transmitted from two satellites. To gain subscribers quickly, the company subsidizes the reception equipment that is offered to consumers at ¥30,000 to ¥40,000 ($255 to $340) instead of the real price of ¥70,000 to ¥80,000 ($600 to $680).

DirecTV Japan entered the CS broadcasting business in 1997. The company was then jointly owned by Hughes Communications, the video rental franchiser Cultural Convenience Club, Matsushita Electric, and the general trading house Mitsubishi Corp. DirecTV Japan offered 128 channels, about 30 of those exclusive to this service, but around one-third were the same as those transmitted by SkyPerfecTV. The service has found it difficult to attract subscribers. Despite its price leadership strategy, offering subscriptions at 10% to 20% lower than SkyPerfecTV, only 231,000 viewers had subscribed by the end of 1998.

To increase the attractiveness of the service, the company announced a new value pack of 16 TV and 29 audio channels for ¥1,600 ($13.60) per month in early 1999. Furthermore, its multichannel "silver" and "gold" package prices were cut by around 10% each per month. Still, the service proved unprofitable. In March 2000, SkyPerfecTV acquired the company. The 410,000 DirecTV subscribers are gradually being moved to SkyPerfecTV. By September 2000, the single remaining CS broadcaster had 2.36 million subscribers, representing around 5% of Japanese households.

A general problem faced by the digital CS operators is that regular analog TV already offers a sufficient number of channels for the Japanese taste. In particular, the NHK analog satellite offer is extremely popular. Hence, the focus of digital satellite is on providing specific services to specific viewer segments, narrowing down topic and theme. Thus, digital satellite broadcasters are following a specific type of niche strategy with several niches, each of which would be too small for an analog provider to offer. Specific sports, such as wrestling and baseball, are examples of such niches.

11.4.1.5 Digital BS offers

Digital services on broadcast satellites started in December 2000, employing the ISDB-S transmission standard. ISDB-S is Japan's digital version of the analog HiVision satellite HDTV standard.

Available via digital BS are the following:

- Three NHK channels (one of them in HDTV);

- Two WOWOW channels (one of them in HDTV);

- Star channel (simulcast of analog satellite service);

- Five channels of the major terrestrial broadcasters.

There is as yet no new programming available. The digital BS services all simulcast existing HiVision or standard analog broadcasts. With around 2 million analog HDTV households and around 14 million analog satellite subscribers to NHK and WOWOW, there is a large installed base available for conversion to digital BS. The fee for the digital service is the same as for analog satellite (¥2,290 per month). Barriers to fast take-up are the competition from SkyPerfecTV, operating in the DVB-S standard, and the high cost of ISDB-S equipment. Set-top boxes retail for around ¥100,000 ($850), and idTVs cost between ¥200,000 and ¥1.2 million ($1,700 to $10,200). It is, however, expected that set-top-box costs will drop rapidly.

In contrast to SkyPerfecTV, the BS broadcasters do not subsidize the equipment. With only simulcast programming available, there is little incentive for viewers to invest in digital satellite equipment.

NHK and WOWOW claim they had over 1 million households receiving digital BS services after the first month of operation. However, 800,000 of those receive the service via cable. Although it is not stated clearly by the companies, those 800,000 households are likely to be digital cable-TV subscribers who already have digital reception equipment, provided by the cable operator.

There are conflicting views on the competitive relationship between CS and BS broadcasters. One suggests that there will be a division of labor between them. It is argued that CS emphasizes multichannel services to take up niches with small numbers of subscribers, and BS, in contrast, places priority on picture quality. However, the line separating CS from BS is becoming unclear, partly because CS, technologically speaking, is as powerful as BS and partly because CS will be allowed to transmit HDTV signals. If CS broadcasters become potential contenders in the HDTV satellite broadcasting market, the nature of strategic interactions is likely to change. Despite NHK and WOWOW having secured around 14 million subscribers in total, there will be a drastic shift toward more competition in Japan's satellite broadcasting arena.

11.5　Cable TV: Weak but getting stronger

The cable industry in Japan has traditionally been starved of investment and hampered by regulation; hence, it has developed very slowly. Digital technology provides the Japanese with a "second chance," but it is not yet clear if they will be able to seize this opportunity.

11.5.1 Slow development of analog cable services

By 1994, cable penetration (cable households verses total households) had only just reached 5%. One reason for this is that the cable industry is very fragmented, having several hundred local cable operators, which renders most operations inefficient. In March 1998, only 139 out of 720 operators were profitable. This is particularly surprising in view of the fact that laying cable infrastructure in Japan can be done at considerably lower cost than elsewhere in the world. In Japan, cable lines are strung from telephone poles rather than dug into pipes laid underground. Furthermore, Japan is a densely populated country so that distances are short. It is estimated that the cost of building a cable network in Japan is roughly a quarter of that in Europe and a third of that in the United States.

However, cable TV in Japan has encountered problems in distinguishing its services from terrestrial and satellite as these already offer a range of channels. The opportunity for the cable industry lies in the ability to offer additional communications services, such as telephony, multimedia, interactive services (for example, karaoke and gambling, which are very popular with Japanese), or video on demand.

In 1994, the government announced a policy shift and recognized the economic potential of cable as an integrator of broadcasting, telephony, and multimedia services. Since then, cable operators have been permitted to expand their operations beyond local markets. The formation of multiservice operators (MSOs) became possible, and foreign ownership of up to 33% is allowed. This has brought in foreign investors, often from the United States, including Time Warner and TCI. Incentives such as low-interest loans from the government aimed to develop the necessary infrastructure.

The introduction of satellite broadcasting in Japan has further stimulated the cable market by creating demand for more than just the usual selection of channels. Moreover, many consumers prefer the simplicity of cable, which enables them to view channels of different satellite providers without having to set up satellite dishes and other system equipment.

Cable TV began to gradually expand its service areas, particularly in major cities. In 1997, cable providers began offering Internet connections and telephone services. By the end of 1999, cable penetration had risen to 21% (i.e., 9.5 million households) but only around 1 million subscribed to cable TV. The remainder used cable for telephony or high-speed Internet.

11.5.2 Digital cable services

While the schedules for digitization of satellite and terrestrial broadcasting have been drawn up, the cable operators have been slow to react. As for the terrestrial broadcasters, digitization of cable requires significant investments, burdening the operators. About ¥1 billion per operator is estimated. To support cable operators, the MPT eased the rules on shared use of broadcasting facilities by multiple cable-TV stations, permitting cost savings. The cable industry is also coming under significant pressure to consolidate and reorganize so as to improve operating efficiency.

The largest cable operator, Jupiter Telecommunications, owned by Sumitomo, Liberty Media, and Microsoft, started digital services in September 1999. The existing 40 analog channels are complemented by 10 digital channels. The digital package costs viewers ¥1,000 to ¥1,500 ($8.50 to $13) per month. In mid-2000, Jupiter merged with the second largest cable company, Titus, increasing its subscriber base to 800,000 cable-TV customers, as well as 142,000 high-speed Internet and 73,000 telephone users as of December 2000. No information is made available how many of the cable-TV customers receive digital cable services. The company's aim is to acquire further cable franchises and expand and upgrade its cable network. An initial public offering is to provide the finances for that.

The second-largest cable operator emerged during 2000 with the merger of the cable-TV businesses of Fujitsu, Secom, Tokyo Electric Power, Tomen, and Marubeni. The new company has a subscriber base of around 500,000 households. It plans to build new facilities for digital broadcast and high-speed Internet access services.

In summary, it can be stated that the Japanese cable industry, as in the United Kingdom, France, and the United States, is late to develop digital services. It remains to be seen if it will be able to use its competitive advantages, such as offering multimedia services, to catch up in terms of subscriber numbers.

11.6 Summary

Penetration of digital services, which started in 1996, is today still low. At the end of 2000, Japan had around 3.5 million digital households, representing 7.7% of total households. Figure 11.1 illustrates the key developments in the take-up of digital TV.

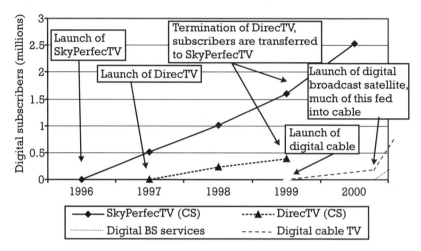

Figure 11.1   Take-up of digital-TV services in Japan.

The merging of BS and CS platforms is likely to have a positive effect on the uptake of SkyPerfecTV services. This is the only service offering true added value while at the same time keeping costs to consumers low. The BS broadcasters have the advantage of a large installed base of analog viewers; however, they have as yet to provide significant added value to viewers to convince them to buy costly digital-TV sets or set-top boxes.

11.6.1 A failed first mover strategy

Japan's HDTV story is an example of a failed first-mover strategy. In the attempt to set a world standard, the country hastily introduced its new technology in the hope of quickly acquiring an installed base, thereby convincing other countries to follow suit. For industrial policy as well as technological reasons, the Europeans, and later the United States, too, refused the Japanese system. In Japan, the HiVision/MUSE standard was introduced as a government-mandated standard. However, this strategy was pursued without support from broadcasters and consumers. The new system proved very expensive and did not offer sufficient improvements over the existing system.

It has been difficult to overcome the NHK inertia to keep the HiVision/MUSE standard. NHK had spent billions on the system and thus strongly opposed the conversion of broadcast satellite to the digital format. The Japanese government had also invested in the MUSE standard and thus has been slow in jumping on the digital bandwagon. Only in

1997 did the government decide that the next generation of broadcast satellites would be digital, and all of the major commercial broadcasters have been given space on the new BS-4 satellite launched in 2000 to operate digital HDTV channels.

11.6.2 Benefits only for the consumer-electronics industry

Japan has 45 million households, and each one owns on average more than two TV sets. For set manufacturers, digitalization implies a huge market: Up to 100 million analog TV sets will have to be replaced by 2010, when analog broadcasting is to be switched off.

For broadcasters, the prospect of digital programming is not so promising, with viewing times of only three to four hours per day. The viewer focus is on specific trendy drama programs, and there is less demand for large multichannel offers. This explains why multichannel via cable and CS digital has shown slow growth. Within five years, CS broadcasting attracted a mere 2.5 million subscribers.

The interest in Japan moving to digital lies not with the broadcasters or consumers but with consumer electronics manufacturers. This industry group has significant influence on government policy making. The country has prospered through the export of consumer electronics products such as TV sets and video recorders. Promoting digital TV is only natural if Japan is to maintain its vanguard position in the world TV market.

With the lobbying power from electronics manufacturers and the industrial policy objective to keep up with international technological developments, the government takes a proactive role in supporting DTT technology. It is the government that is now pressuring broadcasters into digital TV with a "stick and carrot" strategy—and with heavy subsidization. However, for consumers and broadcasters there is no great advantage in switching to digital.

11.6.3 A dilemma for the Japanese government

The danger is that if consumers cannot be persuaded to buy into the new technology, current efforts might yield yet another costly failure. This puts the Japanese government in a difficult position. On one hand, it wants to support consumer electronics manufacturers to keep their position in world markets. On the other hand, it cannot burden broadcasters with the high costs of digitalization. This is why the government pushed back the introduction of DTT into 2003 and is now likely to bear the immense transition costs, estimated at ¥85.2 billion ($725 million). In addition, it has to decide what to do about the HiVision viewers who have bought expensive sets that will become useless once the switch to digital is accomplished.

In comparison to Europe and the United States, it is obvious that technology development and implementation are still very much in the hands of the government. The MPT attempts to reconcile the interests of the different industry players while pursuing its own objectives in the global technology standardization race.

With regard to standard development, Japan has clearly fallen behind. The ISDB standard is employed only in Japan and forms a near-copy of the DVB-T specification. Furthermore, both Japan and the United States have only standardized DTT but not digital satellite and cable.

CHAPTER

12

Contents

Final assessment and conclusion

12.1 European standardization has proven a success

The fact that the digital-TV specifications developed by the DVB Group are being widely employed not only in Europe but also in around 70 countries all over the world speaks for itself. The establishment of a set of common standards for the transmission of digital-TV signals can be seen as a large success. Within only a few years this technology is being used in more than 13 million households all over Europe (see Table 12.1). This represents a penetration rate of around 9%. Moreover, several non-European countries, such as Australia, India, and Singapore, have chosen the DVB specifications in preference to other technological specifications such as the American 8VSB standard.

The DVB standard-setting initiative combined several of the standardization approaches discussed in Chapter 3. DVB is based on a committee process. The agreed-on standards are formally mandated by European authorities; hence, there is an

Table 12.1
Digital TV Penetration in Selected Countries (2000)

	Total TV Households (millions)	Digital-TV Households (millions)	Digital-TV Penetration (%)
United Kingdom	25	6.16	24.6
France	21.5	2.83	13.2
Germany	34.2	1.75	5.1
Western Europe	147.3	13.0	8.8
United States	103	21.6	21.0
Japan	45	3.5	7.8

element of a direct regulatory approach involved. On the issue of CA, a market-driven approach has evolved allowing dominant players to temporarily exert market power. This hybrid approach to standardization, incorporating the positive aspects of each approach, has led to a highly successful outcome.

Europe is currently the most advanced region in terms of digital-TV technology and standardization. In the United States and Japan, broadcasters are still reluctant to embrace the standards mandated there. The U.S. standard for DTT is under criticism from broadcasters for its performance, especially for portable reception. The Japanese DTT standard, almost identical to the European DVB-T, has not yet been employed due to the high costs faced by broadcasters.

Furthermore, the United States and Japan have only standardized specifications for DTT, but not (yet) for DST and DCT. In the United States, this has led to the deployment of several different transmission standards. Hence, it will be more difficult ex post to arrive at a single standard since no operator will want to give up his or her proprietary system. In Europe, common transmission standards for all types of distribution as well as for many additional components have been achieved prior to the launch of services. This has created certainty for companies and a greater preparedness to invest in the development of these new markets.

12.2   Commercial orientation sets the scene for rapid penetration

The commercial orientation of the DVB project has proven to be a crucial factor for success. Consumers will only adopt a new technology if it offers significant added value over existing systems. Digital TV in Europe, which focused on mmultichannel TV and customized services, can offer this. In the United States, by comparison, consumers in most markets already have a large number of channels available. Thus, it was believed that HDTV would be the feature to focus on when introducing digital TV. However, it turned out that HDTV, as pushed in the United States, appears unattractive in particular to broadcasters, but also to consumers: HDTV requires more bandwidth, new production equipment, and new, costly TV sets. At the same time, it does not yield new revenues for broadcasters, which have to carry the additional costs; consumers do not pay for improved picture quality.

The introduction of European-wide standards has led to increased competition, not only within national markets, but also across borders. Since the introduction of digital TV, Europe has seen an array of cross-border activities. The large players have taken the lead but smaller broadcasters as well as public broadcasters are following. Moreover, specific types of content such as international sports events, music, and U.S. movies find audiences all over Europe and can foster cross-border operations. Therefore, although broadcast markets will certainly retain national characteristics based on cultural and language differences, there is also scope for common European content and a common market in broadcasting. The introduction of digital TV can be seen as a driver toward a common European market for both broadcasting and new services such as Internet and e-commerce. Hence, Europe has succeeded in ensuring a major role in the shaping of those high-growth markets that are essential to the emerging information society. This has been the initial intention behind the standardization efforts at the European level.

12.3   Existing market structures impact technology diffusion

When aiming to introduce a new technology, it is essential to consider existing market conditions. This is at least as important as the standardization process itself. Prevailing market structures have an essential impact on the take-up and penetration rate. The differing outcomes in Germany, France, and the United Kingdom described in Chapters 6 to 9 illustrate this. In particular, existing penetration levels of multichannel and pay TV, but also the reputation and brand of existing broadcasters and the regulatory framework, are important factors to be considered by players wanting to introduce digital TV. Business strategies and consumer propositions have to be adapted to those existing structures to make sure that consumers perceive an added value from switching to digital TV.

12.4   Reviving consumer-electronics markets

A further objective of digital-TV standardization has been the revival of the European consumer-electronics industry. According to a report by market research firm Strategy Analytics, European set-top-box production has grown significantly over the past few years to a total in excess of 10.5 million units in 1998, 18.2 million in 1999, and 27.5 million units in 2000. For 2001, the volume is predicted to reach 34.2 million units. This indicates a lower growth rate that is the result of analog pay-TV customers having been converted to digital. However, growth rates of up to 50% are a remarkable achievement in light of the fact that TV-set markets have been stagnant for so long.

The nonstandardization of CA systems is likely to have contributed to this rapid success as it provided the necessary incentives for broadcasters. The fact that DVB has found it difficult to agree on a common API suggests that, here again, nonstandardization may be favorable in the shorter term, especially since this technology is still in the development phase.

12.5   Nonstandardization of conditional access has not led to market distortion

The analysis of the digital-TV developments in three European countries has shown that consumer preferences within national markets are not homogeneous but that there is a considerable demand for variety in terms of content and services. With regard to the theoretical discussion in Chapter 2, this means that indirect network externalities are low, while direct externalities were shown not to apply to TV markets. The developments illustrate that the danger of lock-in and the establishment of a dominant position are small. Each national market for digital TV has at least two competing operators. Although the incumbent pay-TV operators are the dominant players in digital pay TV, this does not mean they will be able to gain a monopoly position in the long term. None of them has succeeded in locking in consumers. Instead, they have provided penetration incentives in the form of equipment subsidization, which has supported the take-up of new services. Furthermore, competition has led to prices being kept low and services improving continually. This also means that there is no imminent danger of vertical foreclosure.

Therefore, the provisions for standardization of transmission and CA technologies as laid down in the 1995 European Directive, and in connection with different national interpretations of these provisions, have shown to be sufficient to ensure competitive markets for digital-TV services.

The fears expressed by regulators and public broadcasters that proprietary CA systems would lead to a fragmentation of the European market and to national monopolies has proved unfounded. Although there are some dominant players, competition exists in each national market and can be expected to intensify further through cross-border operations. Proprietary CA systems have not led to the lock-in of consumers through decoder purchases and pay-TV operators are competing mainly on the basis of content offers. As there appears to be sufficient preference differentiation among consumers with regard to content, pricing, and distribution medium, it can be expected that each market will be able to sustain more than one digital pay-TV operator.

12.6 Minimal intervention by authorities is the preferable approach

The overall conclusion of this book is that rapid diffusion of a new technology requiring high initial investments and coordination is best supported by a liberal regulatory framework. Intervention at an early stage carries the danger that market developments will be distorted or that an inefficient technology will be adopted. The HD-MAC case as well as the standardization developments in Japan show that public authorities are unsuited to pick a winning technology. HD-MAC has been abandoned in Europe. Japan is being held back by its legacy HiVision system.

In network markets where it is unclear how strong network externalities are, it is preferable to support an industry-led approach even if this may involve a temporary period of monopoly or oligopoly power. Once technology diffusion gains momentum, regulators have to ensure that a dominant position is not abused, markets are not foreclosed, and consumer interests are safeguarded. In the case of digital TV, this task falls to competition authorities and media regulators. On a national, and increasingly on a European and global level, media pluralism must be ensured and content control needs to be imposed. National and European authorities have shown in several initiatives that they are taking up the challenges that lie ahead.

The task at hand is far from easy because EU and national regulators are facing a dilemma: On the one hand, they need to make sure that no dominant position is abused; on the other, they are interested in strong European players in media as well as in the consumer-electronics industries so as to stay competitive internationally. Furthermore, regulatory structures need to be adapted to cope with converging ICT industries and increasing cross-border activities.

In the area of advanced television, the EC appears to have learned from the past and has identified its adequate role in supporting European progress in ICT markets. The approach has been not to pursue an interventionist technology policy but to leave technology development and standardization to the market while closely monitoring the processes. Focus is instead placed on competition policy. Here, again, the approach has principally been laissez-faire: Competition authorities on the national and European levels showed their intent of not intervening ex ante in the market. This is illustrated, for example, in the fact that European legislation has been implemented into German and French national legislation without detailed interpretation. At the same time, however, competition authorities are assuming the role of observer with strong powers. They have proven (for example, in the case of the planned cooperation between Kirch, Bertelsmann, and Deutsche Telekom in Germany) that they are able to act—and will do so—should effective competition be endangered.

Glossary

8VSB   8 vestigial sideband

ACATS   Advisory Committee on Advanced Television Services

ADSL   asymmetric digital subscriber line

API   application programming interface

ART   Autorité de Régulation des Télécommunications

ATSC   Advanced Television Standards Committee

ATV   advanced television

ATVEF   Advanced Television Enhancement Forum

BBC   British Broadcasting Corporation

BDB   British Digital Broadcasting

BIB   British Interactive Broadcasting

BMPT   Bundesministerium für Post und Telekommunikation (Federal Ministry for Post and Telecommunications)

BS   broadcast satellite

BSB   British Satellite Broadcasting

BSI   British Standards Institute

BT   British Telecom

CA   conditional access

CCIR   Consultative Committee for International Radio

CD compact disc

CEA Consumer Electronics Association

CENELEC Comité Européen de Normalisation Électrotechnique

CGV Compagnie Générale de Videocommunications

CI common interface

CLT Compagnie Luxembourgoise de Télédiffusion

CoC code of conduct

CS communication satellite

CSA Conseil Supérieur de l'Audiovisuel

CSN Canal Satellite Numérique

CWC cable and wireless communications

DAB digital audio broadcasting

DAVIC Digital Audiovisual Council

DCMS Department of Culture, Media, and Sports

DCT digital cable TV

DF1 Deutsches Fernsehen 1

DNH Department of National Heritage

DSF Deutsches Sportfernsehen

DSK Dvorak simplified keyboard

DST digital satellite TV

DTH direct-to-home

DTI Department of Trade and Industry

DTN Digital Television Network

DTT digital terrestrial TV

DVB digital video broadcasting

EC European Commission

EDF Electricité de France

ELG   European Launching Group

EPG   electronic program guide

ETSI   European Telecommunications Standards Institute

EU   European Union

EU95   Eureka-95

FCC   Federal Communications Commission

FTC   France Télécom Cable

FUN   Free Universe Network

GA   Grand Alliance

GI   General Instruments

HD-MAC   high-definition multiplexed analog component

HDTV   high-definition television

HTML   hypertext markup language

IBA   independent broadcasting authority

ICT   information and communication technology

idTV   integrated digital television

ISDB   integrated services digital broadcasting

ISO   International Standards Organization

ITC   Independent Television Commission

ITU   International Telecommunication Union

KEK   Kommission zur Ermittlung der Konzentration (commission for the investigation into concentration)

LMA   Landesmedienanstalt (state media authority)

LNB   low-noise block

MAC   multiplex analog component

MCR   minimum carriage requirement

MIT   Massachusetts Institute of Technology

MHP   multimedia home platform

MMBG   Multimedia-Betriebsgesellschaft

MoU   memorandum of understanding

MPEG   Motion Pictures Expert Group

MPT   Ministry for Post and Telecommunications

MSG   Media Service GmbH

MSO   multiservice operator

MUSE   multiple sub-Nyquist sampling Encoding

NHK   Nippon Hoso Kyokai (Japanese Broadcasting Corporation)

NTSC   National Television System Committee

NVoD   near video on demand

OFCOM   Office for Communications

OFDM   orthogonal frequency division multiplex

OFT   Office for Fair Trading

OFTEL   Office for Telecommunications

PAL   phase alternation line

PBS   Public Broadcasting Service

PTO   public telecommunications operator

R&D   research and development

SDN   S4C digital networks

SECAM   Système Electronique Couleur avec Mémoire

SES   Société Européenne des Satellites

SMATV   satellite master antenna system

SMS   subscriber management system

TDF   Télédiffusion de France

TPS   Télévision par Satellite

UPC   United Pan-Europe Communications

UN&M United News & Media

USSB United States Satellite Broadcasting Company

VCR videocassette recorder

VoD video on demand

VSB vestigial sideband

About the Author

Katharina Grimme was born in Germany. She holds a diploma in economics and commerce from the University of Wuppertal, Germany, and was awarded an M.B.A. by the Birmingham Business School in the United Kingdom, where she specialized in European technology policy for information technologies. During an internship at the EU Commission's DG III, she worked on the coordination of European policy for research and technological development. She was awarded a Ph.D. by the University of Sussex in 2000 for her thesis on the standardization and technology penetration of digital television in Europe, which is the basis for this book. Dr. Grimme has been working as a specialist in the digital-TV and telecommunications markets for Vision Consultancy, United Kingdom, and for the think tank Prométhée in France. She is currently a senior analyst for digital media at the global research and consulting firm Ovum Ltd., where she researches interactive services, electronic commerce, and the strategies of consumer portals.

Index